Valuing Energy for Global Needs

Valuing Energy for Global Needs

A Systems Approach

Daniel M. Martínez and Ben W. Ebenhack

MOMENTUM PRESS
ENGINEERING

Valuing Energy for Global Needs: A Systems Approach

First published in 2016 by
Momentum Press, LLC
222 East 46th Street, New York, NY 10017
www.momentumpress.net

ISBN-13: 978-1-60650-263-1 (print)
ISBN-13: 978-1-60650-265-5 (e-book)

Momentum Press Environmental Engineering Collection

DOI: 10.5643/9781606502655

Cover and interior design by S4Carlisle Publishing Services Private Ltd., Chennai, India

10 9 8 7 6 5 4 3 2 1

Printed in the United States of America.

Abstract

This book serves as a starting point for energy engineers, sustainability managers, political leaders, and properly informed citizens to explore the net value added by energy systems. Since some resources deplete and some new technologies will require time to emerge, the book takes the reader through the range of costs and benefits, considering the contexts of geography, human needs, and of time. The book takes a particularly close look at the underdeveloped world that currently lacks access to modern energy, and which is crippled by its dependence on dirty, inefficient biomass fuels to meet bare subsistence needs. The authors provide evidence for the reality that energy provides tremendous social value, ranging from the most basic survival to development, to great luxury, inevitably, at a cost. Based on this evidence the reader will be well-equipped to ask the questions: Which energy resources should be abandoned and which should be embraced as we strive for a sustainable future?

Keywords

sustainable energy; sustainable development; energy; sustainability; development; environmental sustainability; energy policy; energy transitions; energy access; energy resources; energy supply; environment; environmental; Developing World; environmental science; oil & gas; shale plays; energy systems; petroleum engineering; energy analysis; pollution; energy science

Contents

Preface

Energy adds value to our lives. Indeed, our very survival depends on using energy to prepare food, sanitize water, and provide heat to our living and working spaces. Beyond that, energy is required for everything we make or do. Its value to us is high: almost like food, water, and shelter. As with all human activities, though, the value comes at a cost, or rather several costs. Of course, there is the price we pay. However, there are also environmental costs and social disruptions with our activities. How do we compare the costs and benefits of energy systems so that we get the greatest net value from our energy choices: now and in the future?

World societies are beginning to embark on substantial energy transitions. If the journey is indeed one of sustainability, the destination should enhance the value we get from our resources, while not diminishing our potential future values. In order to maximize, or at least optimize, value over time, we need means to assess the net value being provided by various energy systems: based on the costs and benefits of each aspect of the acquisition, conversion, and use of each energy source. In order to do that effectively, we must, first, understand what we value and what we seek to enhance by our use of energy.

While it may seem a hopeless task to find common values in a diverse, even polarized, global society, the disparities in views are more based on differences in assumptions than in actual differences in what people want. Fundamentally, we want to be able to provide for our basic needs, our safety and security, our comfort, and ultimately our happiness and satisfaction. What contributes most to those goals is not obvious, but it is relatively universal and assessable.

Some concepts we view as values actually are goals that many of us believe are strongly correlated to the values mentioned above. For instance, economic growth is a commonly held pseudo value. But, in fact, it is not growth itself that any of us desire. It is security, comfort, and happiness. One implicit assumption behind the "unlimited growth" paradigm is that there are no limits to needs, that more is always better. Upon further examination, this is a rather depressing belief; it suggests that we can never be satisfied. Of course, innovations create new technologies and devices, which will always spur some new consumption. However, even this does not necessarily mean growth. We have replaced demand for typewriters and slide rules with demand for computers, demand for saddles, and carriages with cars.

People are prone to think that our desires are unlimited. In the middle of the 20th century, the American dream was to own a home. That dream was clear and achievable for many of us. Now, the dream has become more vague: bigger, fancier, more exotic. How does one achieve such a vague dream? Since everything—energy, land, the world, and even our universe—is finite, we cannot expect to achieve unlimited growth. The misplaced desire for growth is an obstacle to sustainability.

Utopian dreams are always impossible to achieve. When we view energy sustainability in terms of abandoning coal, oil, and gas, and nuclear and large-scale hydropower, we doom our dreams to failure. Oil and gas provide far too much value to our society today to be readily replaced. For anything we want to eliminate that provides critical value, we must know what new resources will replace that critical value. You may think that an iron lung is an inelegant and inefficient form of life support, but you cannot afford simply to shut it off before its replacement is ready. It does not suffice to hate everything that currently provides for our needs. We must rationally assess the true values of energy systems and what resources will offer what values in what timeframes.

The value currently provided by our stalwart energy sources is a baseline. Value is typically assessed by measuring a source's energy content and the price set in the market to trade it as a commodity. Indeed, though, the net value is diminished by a range of negative externalities, which can threaten the goals of sustainability. How shall those be accounted for, to arrive at realistic net value targets?

In order to meet sustainability criteria, it is important not to compromise critical resource systems. These include food, water, atmosphere, the environment, the climate, and energy resources. In a sense, the depletion of energy systems is the least problematic because it is possible to substitute with other kinds of energy, whereas it is certainly not possible to substitute with other kinds of water or atmosphere.

Questions still remain about what levels of impacts these systems can absorb. Every human activity impacts other systems. Land has been converted from the support of natural ecosystems to human use for millennia: planting crops; constructing buildings; paving roadways; and mining. So far, we still have enough land, even to consider growing crops just for fuel. How far can that go?

Over time, humanity has poured billions of tons of pollutants into the air and the water. Local capacities to absorb pollution have been exceeded many times over the last few centuries. Switching to less polluting sources has often helped, as has shifting the center of production. Perhaps, at a global level, we are transforming the atmosphere's composition such that it is altering yet another system: Earth's climate. That may be the impact that is a tipping point from which neither we nor the collective Earth systems can recover. Whether that proves to be the case or not, other impacts on other systems cannot be ignored.

The impacts on the resource systems themselves can be crucial as well. Every resource that taps into a stock can deplete that stock by its use. Indeed, the problem with not being renewable has nothing directly to do with impact on other systems, but to the depletion of the energy source itself. We all know that the fossil fuels will be depleted by human use, but excessive dependence on biomass (such as firewood) can deplete the forests from which they are drawn, thus impacting another system by its depletion.

Finite resources need to be supplemented and/or replaced by alternative options. How can we do this? We must consider the value that these systems offer and how those values can be met effectively with resources that are not stock-limited. Of course, the resources that are not stock-limited still have other, tangible limitations. Solar and wind energy, for example, flow constantly, but are not endowed with the natural quality of being available in stocks. That means that humans can only tap into them at the rate at which they flow: they are then flow-limited. The finite nature of the stocks of fuels is part of their value: they can be tapped when and transported to where they are needed. The fossil fuels represent millions of years of accumulating, concentrating, and refining organic fuel materials. Thus, they can also produce vast, intense energy flows. These properties will be a challenge for the alternatives to meet. Raw firewood and charcoal offer the advantages of being transportable and able to provide somewhat intense energy on demand, which keep them supplying the primary energy to a majority of the world's people, but at a terrible cost. Oil, gas, and coal do offer worthwhile contributions to our current energy systems and will clearly be important players; even the transitions to less-polluting and more sustainable systems are underway.

It is important to understand the values we get from energy and the costs and benefits of the various energy systems to plan the transitions effectively. The transition will not be the same in the Developed World and the Developing World. Not only that, it will also differ by regional needs and availability of energy sources. It will be a complex journey, just as our journey to the present set of energy realities has been. To understand the possibilities, the challenges, and the potential solutions, it will be important to quantify or value energy resources in a more uniform way.

The aim of this text is to provide a foundation to evaluate energy options clearly and as objectively as possible. Our energy future will call on us to make deliberate and informed energy choices. We must be able to appreciate the value we gain from energy use and understand the costs of those energy systems we have and of those that may come from new innovations. This work is meant to be a starting point. There will be—and should be—considerable debate about how values are assigned. We try herein to propose a set of value categories. We offer commentary on the costs and benefits to various systems in each value category. Finally, we demonstrate that, even a cursory approach shows some resources and activities to have extremely high costs.

Acknowledgements

Over the course of the last ten years, Ben and Daniel have worked together on several projects, writing and presenting on topics ranging from energy transitions, to energy poverty and accessibility, to incorporating sustainability into undergraduate engineering curriculum. In that time, both Ben and Daniel embarked on new academic positions in petroleum engineering and environmental science and policy, respectively, teaching about energy resource utilization and its impact on economic, environmental, and social structures. Both would like to thank the many colleagues, students, and family members that have helped shape their thoughts and investigations over that time, and especially during the writing of this book. In particular they would like to thank their wives, MJ and Thalia, and their kids. They would also like to thank: Rob Sanford from USM's Department of Environmental Science and Policy for supporting Daniel in his early academic career and pushing him to publish more; colleagues throughout Marietta College's faculty for supporting Ben through the tenure (and writing) process; Joao Chidamoio and Joao Venancio from UEM in Mozambique for their work with AHEAD Energy and their efforts to be good stewards of the resources endowed in their country; the students that helped with early drafts of this book as well as with helping to generate some of the images: Aaron Dillon, Mitch Houston, Michael Lichtnefels, and Jennifer Starkey. Finally, Ben and Daniel would like to thank Momentum Press for their support in bringing this book to market.

Gorham, Maine and Marietta, Ohio, October 2015
Daniel M. Martínez and Ben W. Ebenhack

CHAPTER 1

Energy's Role and Value

From Hunter-Gatherer to Modern Man

For most of our existence, humans have lived as foragers, hunting, gathering, and fishing for survival. However, the end of the last ice age marked an important opportunity for improving our condition. With arable land now abundant, our ancestors deliberately began to cultivate plants and domesticate animals to enhance and secure food production. As a result, society rapidly transitioned from the hunter-gatherer to one dominated by settled agriculture, creating civilization much as we now see it, and with populations larger than ever imaginable.

Once this more energetically secure social structure took hold, and as towns and cities began to emerge, a small subset of people began to pursue other endeavors not related to food production or acquisition. They focused their efforts on art, architecture, and engineering, creating tools for comfort, learning, and warfare. And as these complex social and economic activities developed, ever more external energy was required for accessing water, heat, and materials for cooking, lighting, and manufacturing, respectively (Smil 2012).

First Forays into External Energy

Biomass was humanity's first major external energy source. The acquisition of fire truly set us apart from other species, and that leap forward holds its place in the creation stories of virtually all people around the world. The truth is prehistoric, so we cannot know how humans first came to use fire, let alone who deserves the credit for that breakthrough. Fire may well have been used opportunistically for a long time before people discerned ways to build them. However, once we had complete control of fire making, it changed everything. No longer were we confined to a diurnal existence. Some people were able to work after dark. All were able to eat a much wider range of foods, including other animals, with a significantly reduced risk of disease.

Human invention did not stop there. Some thousands of years ago, people found that by containing the fires within certain enclosures, it could be concentrated into rather intense heat. This allowed for the processing of certain stones to yield metals that could be worked to make a huge array of tools. Some stones, primarily clays, could be baked to make durable, even artistic, vessels. And as material processing improved through the centuries, raw metals such as copper would be replaced by stronger bronze amalgams, which, in turn, would be replaced by even stronger iron and steel. These transitions were huge steps in the rise of civilization. (However, with lack of civility, the new materials were often put to use in warring on other people.) However, raw biomass fires could only take us so far.

Nonthermal external energy sources were also tapped to ease the needs of human and animal labor throughout this great civilizational transition. Water wheels and windmills were employed extensively throughout the world for milling grain and pumping water. Wind was also used for propelling ships across small and great expanses, primarily for exploration, extraction, and trade of other precious resources. While humans were still used heavily for many precise and laborious tasks during this time, the trend of increased external energy use, coupled to increased mechanization, utterly transformed the human condition (Smil 2012).

Indeed, every step of progress has demanded more energy. The firewood and charcoal that had moved humanity so far beyond the other animals with whom we had competed for millions of years had finite growth rates, which could be exceeded. In Europe, for example, extensive wood use resulted in drawing down the stocks of the once abundant forests. By the 17th century, development had largely deforested the British Isles near population centers (Perlin 2005). Thus, as wood became scarce, the search for new external energy alternatives gained impetus. And yes, fossil fuels would have been considered alternatives, especially when compared with raw biomass use of earlier times—and of present times for nearly half of humanity.

Coal for Thermal Applications

Coal was the first major alternative external energy source. For centuries, coal was known to burn similarly to biomass, but received little attention because it produced a great deal of unpleasant smoke when burned in the same manner as wood. However, as wood became scarcer and more precious with the decimation of forests, experiments with coal showed that it could burn more cleanly and more intensely than wood if done so under certain, well-controlled conditions. And with that technical breakthrough, coal could now serve the same thermal purposes as the dwindling biomass combustion resource.

Another benefit of using coal was that its energy content was much greater than that of firewood: about twice as much. From a utilitarian perspective, that meant a wagon load of coal contained twice as much, or more, energy per haul. Moreover, coal, being the altered remains of many millions of years of plant growth, was readily found concentrated and compressed in finite coal seams, which could yield the equivalent of many acres of felled trees in just one day's digging. Its thermal value was evident: coal could provide more intense heat than firewood and could be readily extracted and moved to the economic centers where intense energy consumption occurred.

Indeed, the sustainability of the early Industrial Revolution, powered by steam engines made of forged amalgams of the various metals, depended on the thermal energy value added by coal. So did the survival of the British forests. Coal was able to displace much of the demand for firewood and charcoal, while meeting exponentially growing demand for energy. By the year 1800, coal was producing as much energy in London as the sustainable growth of forests the size of the contiguous United States. Seemingly ironically, Britain's ecosystem sustainability was enhanced by fossil fuel use in that era.

Commercial coal mining quickly produced energy that dwarfed fuelwood production, while providing energy to support tremendous development. But, the benefits of coal came at a high cost. Mining was an extremely dangerous activity. If it was not mine collapse, it could be: explosions, asphyxiation, fires, or black lung disease. And, sadly, the wealth procured by mining companies was not shared effectively with the workers whose lives were imperiled in obtaining the resource.

Petroleum for Lighting (and More)

The energy transition to petroleum also was in response to the limited carrying capacity of another "renewable" energy source used heavily as a lighting fuel: whale oil. Although it may seem callous to talk of them as such, whales would have been categorized as a renewable resource, as they do reproduce on a relatively short and observable time scale and have a relatively high biomass chemical content. Nevertheless, when whales were being hunted as an energy source, human use was depleting the biotic stock much too rapidly, thereby threatening an ecosystem and a sentient species. (We see here a poignant example that renewability does not always equate to environmental goodness.) Fortunately, the pinnacle of whale oil use was short-lived. It was soon replaced by a resource that offered superior value, with relatively fewer environmental externalities, again, for that time period.

Prior to its use as a whale oil replacement, petroleum had been known and utilized for quite some time. In the "New World" it was often referred to as Seneca Oil, because the Seneca Indians were seen skimming it off creeks and using it as a "crude" fuel, as a sealant, and even as an elixir. Indeed, nonfuel product uses preceded extensive energy production from petroleum. However, when experimentation revealed that a fuel could be readily distilled from crude oil that burned cleaner and brighter in lamps than whale oil, a new industry was born. A modest investment led to the first well that produced more oil than a dozen whaling ships. The "black gold" gushed out of the ground, yielding vast, dense flows of energy.

The production of petroleum and its sister product, natural gas, often swelled beyond demand, causing the commodity prices to crash. This boom and bust was actually a significant factor in the value added by oil. The supplies were able to meet and outstrip demand, allowing new uses to grow up around the cheap, abundant energy. The Industrial Age itself transformed radically with the new energy alternative, with mobility being a large part of this transformation. Locomotives and steam ships had evolved with the use of coal, and with firewood in North America, thanks to vast tracts of virgin forest. However, petroleum products offered denser and more readily controlled energy.

Daniel Yergin's popular tome, *The Prize*, focuses extensively on the role that petroleum played in the 20th century, particularly in warfare (Yergin 2011). While coal-fired steamships had a great advantage over sailing ships in battle, oil-fueled ships were vastly superior again. Of course, petroleum fuels also permitted new, faster, and more powerful forms of armored automotive and aviation cavalry, and aerial combat, in particular, would have been impossible before the energy dense, liquid fuels. Indeed, modern society, even beyond the foolhardy needs of warfare, would not exist as we know it without the advantages of these new alternative fuels.

Natural Gas

Natural gas has had a history of replacing coal as the preferred fuel in a number of applications. For many years, and before electric lighting, methane and other flammable gases were distilled as by-product from airtight bituminous coal ovens and distributed throughout towns to light streets and homes (Thomas 2015). This manufactured "town gas" was eventually replaced by natural gas in areas where it was discovered in abundance and near town centers. Natural gas, with many fewer impurities than town gas, burned very cleanly, although it was harder to transport long distances or to store.

It was not until the flame technology of innovators like Robert Bunsen that a natural gas industry really developed, as wider applications occurred beyond lighting and into cooking and heating. As the United States and Europe began to switch from the tedious and smoky use of coal for heating and cooking, pipelines were laid to bring this valuable gaseous product into homes and businesses.

Recent technological developments are bringing tremendous new quantities of natural gas to the marketplace, even as forecasts from just a few years ago had predicted quite the opposite. This is due in large part to the shales: tight, impermeable rocks within which the original organic material accumulates and transforms into oil and gas. Shales were thought to be unproducible "source" rocks until very recently. Now, new technologies to control very accurately the path of horizontal drilling are combined with extensive application of hydraulic fracturing technology to liberate enormous quantities of natural gas from their tight sepulchers. As a result, natural gas's value, especially as a coal alternative for electricity generation, has grown even greater in recent years.

Electricity

Electricity is an energy carrier rather than a primary energy source, but it is one that has added fantastic value to society. Not only extremely clean at the point of use, electricity enables a plethora of applications, many of which were unimaginable to past generations. It is constantly available at the touch of a switch on every wall in every room in the Affluent World. We tend to agree with Vaclav Smil when he asserts in his introductory text, *Energy: A Beginner's Guide*, that "no societal transformations would be possible without the conversion of fossil fuels to electricity" (Smil 2012). Both speak to the need for dense, abundant, and transportable energy sources that the fossil fuels provide, as well as the robust and flexible nature of electricity.

For example, as a clean and efficient means to extend the day, electric lighting is unrivaled. The difference between turning a switch to initiate the soft glowing light of a tungsten filament encased within a glass bulb, and the constant refilling of a kerosene lamp whose combustion products and odors envelop a room, must have been striking, if not initially frightening. Either way, it soon took hold. After being introduced commercially in 1880, the electric lamp nearly wiped out the petroleum industry in a very short time period. The Edison Electric Light Company was determined to do just that; however, Henry Ford's mass-produced automobile uncovered a new market from which petroleum companies could thrive.

Coupled with the ability to control the energy very minutely, electricity also powers millions of individual machines, which are responsible for much of the development of the past half century, most recently in the form of ever more important electronics and communications industries. Indeed, the ability to talk and do business all across the globe at all hours of the day on computers, tablets, and mobile phones, has transformed individual national economies into one vast, intertwined global economy. The amount of useful (and sometimes useless) information has never been so readily available and accessible in the history of humanity. It has improved lives and seemingly transformed culture from one where the notion of time had real meaning, to one that expects everything to be available right now, at all hours of the day, without interruption. This could not have occurred without the use of electricity.

Climbing the Development Ladder Through Energy

If we glean nothing else from the history of the human condition, it is that humans naturally trend to higher levels of social needs as they move from bare subsistence to being productive members of a complex society. This often entails the need to use more and more external energy to satisfy growing needs and wants. Moreover, it has been observed that, if given opportunity and affordability, people will naturally adopt new energy systems if the systems will improve the overall quality of life. This new fuel adoption strategy is often referred to as an "energy ladder," though it is probably more accurate to call it an energy web. In these strategies, people will use external energy sources at the lower levels or rungs for cooking, heating, and lighting in order to fulfill basic human needs, then climb up the ladder, using more advanced fuels as they try to meet higher level needs, such as in gaining better health and education, and in the accumulation of wealth (UNDP 2005).

Someone reading this book likely lives in a Developed Country, which may put them out of touch with what it is like to be without the modern amenities that modern energy affords. Thus, looking at Developing Countries allows us better perspective on the transformative nature of modern energy. Let us think about rapidly developing nations like China and India for a moment. At low-income levels, biomass will be the dominant fuel type, because it is easy to acquire and use. As income and social status rise, more modern fuels and electricity will be adopted, though often, traditional fuels are not abandoned altogether. Typically, the more modern fuels will be mated to more advanced technologies, and the fuels will tend to be more versatile, more powerful, and cleaner than traditional fuels. The result of this positive societal progression, and resultant new fuels adoption, often is massive population growth, ever increasing energy consumption to meet new demands, and the resulting degradation of the natural environment, sometimes severe.

Indeed, a hallmark of human activity is combustion, and by proxy, pollution. Whether it be to build a wood fire for warmth, to use a natural gas stove top to boil water, or to burn coal to generate electricity from steam turbines, all require the basic input of hydrocarbon fuel and oxygen from air.[1] Since oxygen is only one component in air and since the hydrocarbon fuels are of different quality, these combustion processes occur at varying degrees of completion. As a result, unpleasant emissions are produced, the most obvious and obtrusive being the smoke that enters our air.

Few things alter Earth's ecological systems as much as the use of energy for human endeavor. We use fuels and machines to extract mineral resources (not the least of which are energy resources), to cultivate food crops and building materials, to transport people, and to manufacture and deliver goods. This inevitably results in the production of unwanted emissions into the air and waters. Unfortunately, with progress comes the consequence of negatively impacting our people, our places, and our planet. And we should be mindful that the scope and magnitude of these consequences are entirely governed by our consumption choices.

To be clear, humanity, in its current state, needs abundant energy to satisfy the voracious needs of our modern, global society, and it is likely to continue to be heavily supplied by the fossil fuels. However, in this new century, we have become keenly aware of the physical, economic, environmental, and social impacts that constrain current energy choices. Public attitudes have changed, especially in the last 50 years, over how we should deal with issues related to environmental protection and sustainability, in general. We live in an era where much of the world has adopted strict rules to protect air and water quality, including gases that may be impacting the global climate. We also live in an era where the idea of universal access to

---

[1] The word "hydrocarbon" is often used in reference to oil and gas, which comprise compounds that are essentially just carbon and hydrogen atoms. Here we are using the term to note that all of the natural combustion resources, from wood to natural gas, consist of carbon and hydrogen based compounds.

energy is finally gaining traction, realizing that energy resource extraction to benefit only half of humanity is more harmful than helpful, if the balance of trade does not include goals of modernization in resource-rich, cash-poor nations (see Khodeli 2009).

Thus, as we continue to see and support more nations climbing the development ladder, we have realized the limitations of relying on the depletable and polluting fossil fuel energy resources to satisfy energy needs of over 7 billion people. Indeed, humanity is beginning to adopt better energy efficient practices and technologies to improve our energy utilization rates (though not nearly as aggressively as we should). We also are beginning to view nondepleting earth- and sun-based energy as the next necessary energy transition, in the form of geothermal energy, hydroelectricity, wind electricity, and the perceived "holy grail," direct solar electricity. The adoption of these "renewable" energy systems into the global economy thus far has progressed in fits and starts, more out of a need to reduce local dependence on foreign energy sources, and to reduce dependence on nuclear power. More recently, it is being aggressively adopted to reduce unwanted carbon dioxide emissions. This pattern of fits and starts likely will continue to be the case throughout this century, because as we move forward, we will realize that fossil fuels comprise an extremely versatile set of characteristics that are suited perfectly to existing infrastructures. This will cause them to continue to hold value in the global energy system and continually challenge alternatives that seek to replace them throughout this century.

The Global Energy System

Now that we have gone through a brief introduction of where we came from, in terms of our energy choices, let us now consider our established energy system.[2] Generally speaking, the global energy system is understood to be a juxtaposition of energy supplies and energy demands. Schematically, the system consists of a number of stages that looks like

---

[2] A more complete introduction to the history (and future) of our energy choices can be found in our 2013 text, *The Path to More Sustainable Energy Systems: How Do We Get There From Here?*

the following simplified chain, which has been modified from Kornelis Blok's *Introduction to Energy Analysis* (Blok 2007):

Acquisition* → *Extraction* → *Conversion* → *Distribution/Transportation* → *End-use Utilization* → *End-use Impacts

The first stage of the energy system is *acquisition*, or identification of the resource to be tapped. Some examples include oil and gas exploration and discovery, wind speed analysis and assessment, and solar insolation mapping. Fossil fuel resources occur underground, so exploration via drilling or digging is required to find accumulations of economic proportions. Geothermal resources similarly utilize underground thermal reservoirs and also require some exploration of the subsurface. The solar resource exists universally across Earth, but not uniformly, dictating the need for site specific acquisition measurements that include an understanding of local climate, weather patterns, geography, and proximity to natural and built structures. Similarly, wind acquisition requires a good understanding of localized conditions, and is very much a function of topography and elevation; thus, a good deal of work goes into detailed mapping before extraction. Hydropower acquisition potential exists wherever there is moving water, but useful energy production is a function of the water flow rate and the vertical drop that water falls through, described by engineers as the hydrostatic "head." Damming the flow of a river concentrates the head, increasing the amount of energy produced by a single extraction facility. Thus, we can manipulate the available acquisition potential at specific sites. Finally, biofuels can be acquired where there is abundant arable land and easy access to water and sunlight.

The second stage of the energy system, *extraction*, refers to tapping or extracting energy from primary sources. This can be: coal mining; oil and gas production; the use of wind or water turbines; or harvesting of raw biomass for nonfood energy use. Mining is sometimes used as a collective term that would include operations through wellbores, that is, drilled holes through which resources can be extracted. We find it useful to distinguish drilling and production through the resulting wellbore from mining operations. We do this, in part, because underground and surface- or strip-mining are likely to have higher environmental and social costs than most activities that can be conducted through long, slender wellbores. Also, oil and natural gas, and geothermal energy are all extracted through wellbores, and extraction through wellbores is also, in general, much more energy efficient than mining solid material. A chief exception to that observation would be mining done for oil from tar sands. The extraction of sun-based energy often overlaps with the conversion stage, as the extraction of solar, wind, and hydro resources goes directly to converting the primary energy into electricity. In the case of engineered biofuels, extraction is tantamount to harvesting a food crop, then processing it biochemically into combustion fuels that can be used similarly to, and often with, oil or gas.

The third stage, *conversion*, is the initial process involved in converting one energy form to a more useful form. This can be: thermomechanical conversion in coal or natural gas power plants to make electricity; refining oil into its many products; or converting a form of kinetic energy, such as the flow of wind or water, into electromechanical energy for electricity generation. The initial conversion is generally to a carrier form. Electricity or portable fuels such as gasoline, diesel, heating oil, and jet fuel are typically the forms that enter the economy. (Natural gas does not require refining, but if it is to be shipped overseas in the next stage, it must be liquefied at extremely low temperatures.) All of these conversions prepare the energy to be delivered to the consumer.

The fourth stage, *transportation and distribution,* moves the energy to the point of consumption. It can be: pipelines carrying natural gas to homes and businesses; the electric power grid; or trucking fuel to service stations. Electricity is commonly produced into a distribution grid. The more extensive the grid, the more it can, itself serve as a storage buffer. Oil and natural gas are generally transported onshore through pipelines, which is the safest and most effective means of moving material, when feasible. Intercontinental transport is done by tankers. Coal is generally transported by rail or barge. Firewood in the Developing World is carried where people live close enough to the forests to gather wood themselves, and this work is commonly performed by women. Wood is transported in trucks and old buses and on the backs of bicycles to urban areas. Energy may be transported to an intermediate destination (such as an electric power plant, a refinery, or a fueling station). It is always delivered to the end point of use.

The fifth stage, *end-use utilization*, is the stage of finally producing useful work to the consumer. It can be in the form of: powering electronics; converting electricity to light or heat; powering an engine; or burning fuels for space or process heat. Raw biomass is almost always burned directly for heat: cooking is the dominant use globally. Electricity has opened the door to an immense array of electronic appliances, while petroleum fuels have found primary use in internal combustion engines to power mass transit and highway systems.

The sixth stage, *end-use impacts*, is the effect of our use. Of course, there are impacts associated with every step in the process, but only at the end-use can negative impacts be contextualized with the benefits, or values, achieved. The scale of operations (i.e., consumption) tends to be tightly linked to the scale of impacts. The fossil fuels, collectively, produce at staggering scales to provide the world with some 84% of total anthropogenic energy use. It can, then, not be surprising that the fossil fuels also produce a significant majority of the environmental impacts. Some resources will probably be better: solar, wind, geothermal, and some hydropower projects will likely produce energy with less impact per unit of production. It is foolish, though, to assume that any resource will be totally impact-free, and it will be difficult to know the extent of impacts as contemporary alternative energy systems that have been producing at extraordinarily small scales ramp up to scales that can compare with those of the fossil fuels. Some resources that are used extensively, particularly traditional biomass, have demonstrably worse impacts than even the worst of the fossil fuels.

Key Energy Demand Sectors

The six stages of the energy system described above lay the foundation for the current economy of much of the world, more or less broken down into four key economic sectors: *electricity generation*; *transportation*; *industry and manufacturing*; and *residential and commercial end-use*.

The electricity generation sector often is the largest consumer of primary energy in Developed Nations, and influences all of the other economic sectors except for transportation, at least, directly. As such, electricity generation is distinguished because it is not an end-use sector of the economy, but a necessary energy service sector, for which the rest of the (nontransportation) economy depends. As an example, the United States Lawrence Livermore National Laboratory (LLNL, 2014) reports that total national primary energy consumption in the United Sates in the year 2013 amounted to 97.4 quadrillion BTUs, or Quads for short.[3] Of that, 38.2 Quads, or 39% of total consumption, was diverted to electricity generation. Coal, natural gas, and nuclear fuels represent the majority (86.6%) of primary energy used for generation. (The primary energy sources that fed into the electricity sector are broken down in Table 1.1.)

Note that of the 38.2 Quads entering the electricity sector, only 12.4 Quads actually flowed into the other demand sectors, with 25.8 Quads leaving as rejected, or waste energy. The reason is: there is a high cost associated with electrical energy conversion efficiency, as much of the primary energy conversion processes involve combustion, resulting in heat losses that cannot be tapped. This is often referred to as being "lost out the stack" although significant thermal pollution into nearby rivers and

Table 1.1 Energy sources fed into the U.S. electricity generation sector, 2013.

Energy Source	Energy Fed into Sector (Quads)	Total (%)
Coal	16.5	43.2
Natural gas	8.34	21.8
Nuclear fuels	8.27	21.6
Hydropower	2.53	6.6
Wind	1.59	4.2
Biomass	0.47	1.2
Petroleum	0.26	0.7
Geothermal	0.16	0.4
Solar	0.08	0.2

Source: LLNL 2014

[3] A Quad is such a large number (one with 15 zeroes after it) that it is rather difficult to conceptualize. Syracuse University's Maxwell School of Citizenship and Public Affairs (2014) offers a little less abstract analogy of it being equal to 45 million tons of coal, or a pile of coal that is 10 feet tall, 1 mile wide, and 3.3 miles long. By their estimates, it would take 9 minutes to drive around it, if you traveled at a speed of 60 miles per hour.

streams also occurs. This reality offers some advantage to hydro-, solar-, and wind-based energy resource systems, at least from a fuel's accounting perspective. A megawatt of hydro, solar, or wind electricity generation already accounts for inefficiencies. That is, we do not bother to discuss the inefficiency of the wind or water turbine, or of the concentrating solar power system at the point of conversion, because we consider water, wind, and solar radiation to be "free" sources of energy, as opposed to coal or natural gas, which are priced in the market before they are converted into electricity. Taking this into account, solar's contribution to the electric power sector actually jumps from 0.2% of energy input for electricity to roughly 0.7% of electricity used, since none of it is rejected. Regardless, the share of energy generated from direct solar radiation remains a dismally small contribution toward our energy system. Even the resulting electricity production from solar energy is barely on par with the effective energy flow from petroleum into the sector, which is only used as a backup fuel when local grids are approaching peak electrical loading. Furthermore, solar and wind remain small contributors to the only sector in which they are directly used.

The transportation sector is another large consumer of primary energy. The scale of its influence on a particular national economy is often a function of the physical size of the country, the size of the national economy, and the primary modes of transportation chosen to move people and goods into, out of, and around it. In the United States, a country that spans a substantial land area, there is a heavy dependence on road and air transport. In fact, in 2013, 27 Quads of primary energy was diverted to its transportation sector, which is about as much total primary energy consumed by all of Central and South America combined. Primary energy and electricity fed into the sector, but the mix is very much lopsided to one primary resource: petroleum. (A breakdown is given in Table 1.2.)

Table 1.2 Energy sources fed into the U.S. transportation sector, 2013.

Energy Source	Energy Fed into Sector (Quads)	Total (%)
Petroleum	24.9	92.2
Biomass	1.24	4.6
Natural gas	0.79	2.9
Electricity	0.03	0.1

Source: LLNL 2014

Petroleum dominates the U.S. transportation sector, as it does every other country. It will be challenging to displace it in this sector, because petroleum products have remarkable energy and power densities. In other words, a relatively small volume or mass of gasoline or diesel provides for long cruising range and rapid acceleration: critical values for transportation. Biomass, almost entirely ethanol from corn crops, is in a distant second place in the United States energy mix, which is a heavily subsidized endeavor. Compressed natural gas may be able to take some noticeable market share, where the shale plays have made gas much more abundant in the United States. For electricity to take a much larger share, batteries or other storage technologies will need to advance dramatically. Liquid fuels will remain valuable to consumers for the foreseeable future.

The industrial/manufacturing sector is yet another large consumer of primary energy. In a massive economy such as the one in the United States, there exists a large demand for fuels, materials, and other manufactured goods. Primary industries include petroleum refining, and steel and aluminum processing, while secondary industries (i.e., industries that depend on the primary industries to function) comprise cement, paper, and chemicals industries. All of these industries require healthy amounts of primary energy and electricity to make them operate; thus, it should not be surprising that manufactured goods contain a large degree of what is called embedded or embodied energy: energy that is required to produce all of those manufactured products. The U.S. industrial sector consumed 24.7 Quads of energy in 2013. (The breakdown of energy sources fed into the sector is shown in Table 1.3.[4])

Table 1.3 Energy sources fed into the U.S. industrial/manufacturing sector, 2013.

Energy Source	Energy Fed into Sector (Quads)	Total (%)
Natural gas	9.08	36.8
Petroleum	8.58	34.7
Electricity	3.26	13.2
Biomass	2.25	9.1
Coal	1.50	6.1

Source: LLNL 2014

[4] Though electricity is an energy carrier, not an energy source, it is often accounted for in energy analysis as a feed source into a sector.

It is obvious from viewing Table 1.3 that natural gas, petroleum, and electricity are vitally important to the U.S. industrial sector, since petroleum refining and chemicals manufacturing are the two dominant industries in the country. Coal's contribution is somewhat obscured by two factors: electricity and China. Since 43% of all electricity generated in the United States comes from coal, the total contribution to the industrial/manufacturing sector will obviously rise if we count the embedded "coal energy" in electricity. Also, much of the metal processing that the United States used to perform, which requires large quantities of coke, now is outsourced to other countries, primarily China. This means they use the coal and produce the emissions, while the rest of the world benefits from just using the steel they produce and sell to us. This is another form of embedded energy, and deferred local pollution, that does not get accounted for in our total consumption tally.

Finally, the residential and commercial sector rounds out the energy system's demand component. Homes and office buildings, combined, represent yet another substantial fraction of energy use in modern economies. These are end-use sectors that depend heavily on electricity, which is already processed from primary energy, and that also depend on natural gas. Common activities, of course, include: lighting, heating, operating appliances, and also communication, and entertainment. As such, the energy sources fed into these sectors are varied, and in the United States, amount to 20 Quads of primary and electrical energy (broken down in Table 1.4).

Table 1.4 Energy sources fed into the U.S. residential/commercial sector, 2013.

Energy Source	Energy Fed into Sector (Quads)	Total (%)
Electricity	9.32	46.6
Natural gas	8.41	42.0
Petroleum	1.37	6.9
Biomass	0.53	2.6
Solar	0.232	1.2
Coal	0.05	0.25
Geothermal	0.04	0.2

Source: LLNL 2014

Electricity and natural gas are the two energy-feed sources that dominate end-use sectors, because they are the cleanest and most flexible of the fuels for lighting, cooking, and space heating. In certain areas of the country, petroleum is still used directly in domestic and commercial buildings as heating oil for space heating. Some firewood and tiny amounts of coal are still used for space heating, even in the United States. Geothermal applications in buildings remain small, and would generally consist of tapping moderate enthalpy heat such as those found in the western part of the country for municipal heating.

Other Countries

The data summarized here use the United States as the example, and while the magnitude of energy used in that country is much greater than most other countries, the distribution of usage in other affluent, industrialized nations would be similar. Many of these other nations would have less of their total energy going to the transportation sector. The American affinity for automotive transport has oft been described as a "love affair," and most other industrial nations have far more extensive public transportation networks. Additionally, the country of France is noteworthy in having its electricity generation sector dominated by nuclear power. Also, in places like China, coal is used heavily as a cheap residential heating source, though that is changing, as the need to divert coal to electricity production increases.

One of the important things to note from Tables 1.1–1.4 is how truly infinitesimal the role is played by the popular renewables in the current energy mix. Although some of them may offer great potential value, even collectively, they do not provide significant value to the current overall energy mix, especially within the current energy infrastructure. This represents a huge challenge if we seek energy systems that offer greater sustainability characteristics.

Characterizing Energy Sources

As we have been alluding to, many people, particularly in industrial and in postindustrial nations, long for rapid transitions to renewable energy sources, as they are perceived to be at the pinnacle of the energy ladder, meeting the sustainability criteria of being abundant, cheap, and non-polluting. However, simple definitions such as "renewable" and "nonrenewable" do not adequately describe the differences between the many primary energy sources, including how their availability is measured, how they are tapped for exploitation, and how they are integrated into the economy. Instead, it is more accurate to describe energy sources by their limitations in how and when they can be used.

Stock- versus Flow-Limited Energy

The best convention that we have seen for describing energy resources comes from Donella Meadows, in her wonderful, posthumously published book, *Thinking in Systems* (Meadows 2008). In it she introduces two terms that we feel better describe the properties of resources: stock-limited and flow- or flux-limited. Let us review them here.

A stock-limited energy resource is one that already has an accumulated amount available in the ground to be tapped, and can be used as soon as it can be extracted, at whatever rate we choose, but whose total extractable amount (i.e., reserve) is finite and can be depleted. Fossil fuels are a good example of this kind of resource. Coal, oil, and gas represent millions to hundreds of millions of years of stored solar energy that exist as deposits within the ground. A key benefit of being able to depend on such resources is that human activity can proceed with a high degree of certainty and predictability because that energy is available in the proverbial energy "bank." Fossil energy stores are both sizable and available to draw from available reserves right now. That is extremely important when trying to power a global economy. A key drawback of our dependence on a stock-limited resource is that since it is finite and depletable, it is gone once we use it, and the faster we extract it from the ground for use, the shorter its lifetime. Thus far, those limits are only beginning to be noticed by humanity. The caveat is that because consumption rates grow exponentially, the "noticing point" is never that far away from a rapid depletion of that resource.

On the other hand, a flow-limited energy resource is one that has no discernible stock, but rather flows from an independent source that can be harvested indefinitely. They actually have no reserves to draw on, which means that our use of them is restricted to the amounts that can be tapped as they flow to and on the Earth's surface, thus the term, flow-limited. These resources do have the advantage that they are not depleted by our use, but they are completely limited by external natural forces, and must always be available to a collector that, in turn, needs to use it immediately. The best example of this kind of energy source is the sun. Solar radiation constantly bombards Earth as a whole at a discernible rate. Practically speaking, humans will never have to worry about running out of solar energy. What they do have to worry about is all the physical obstacles that impinge, divert, or diffuse the solar energy flow from our Earth-bound collectors. When the sun is not shining because it is night time, or is obstructed by cloud cover, the flow rate can be reduced to virtually nothing, making the resource useless without some artificial means of storage. It is difficult to imagine running our massive global economy with that kind of resource, given existing infrastructures.

In terms of long-term availability of a given resource, flow-limited energy sources will be vastly superior to stock-limited resources. The sun will last for a few billion years and in the meantime, it does provide a great deal of energy to Earth's surface. But, it is clear that while the flow of energy from the sun is vast, it is limited. Thus, talking about the virtually limitless energy resource from the sun (as many solar advocates do) is actually meaningless to humanity in a practical sense, because the limitations on the rate and intensity of flow are significant. Indeed, the amount of solar radiation reaching any point on Earth's surface is small. At a typical average 250 W/m^2, the solar flux is the equivalent of the energy consumed by a small chandelier. Upon considering the efficiency of conversion, a square meter of surface would power a single compact fluorescent fixture. In contrast, the production from a single natural gas well, in its first year, would power more than 10 million such bulbs for some 20 years. The differences are that solar energy is found spread out over Earth (although certainly not evenly distributed) and will go on for billions of years rather than a mere tens of years.

Clarifying Issues of Environmental Impact

The solar resource, as well as those resources derived from the sun, does have the advantage that they are not depleted by our use. They are likely to be cleaner than the combustion or nuclear fuels. However, we must be able to distinguish between the merits of availability for the future, versus environmental impacts.

To do so, we challenge the notion that renewal rates have anything to do with the environmental quality of a resource. We have already introduced the absurdity of considering whale oil to be an environmentally sound resource, in spite of the fact that whales are living things that reproduce in a reasonable human time frame. Thus, too, we suggest, rather, that the appeal of resources that are limited by their flux (how fast they flow to or on Earth) rather than by their stocks is an economic issue. That is, how long will be able to tap the resource? If we only tap the natural flux of solar or wind, for instance, our use will never deplete them.

Hydropower is considerably more concentrated. Indeed, it combines some of the advantages and disadvantages of both stock- and flow-limited resources. Damming the flow of streams or rivers further concentrates the density of the energy flow and creates a stock of potential energy in the height of the water behind the dam, which can be tapped on demand, rather than on being dependent on the momentary fluctuations of solar energy. The dam, though, not only concentrates the potential energy of water captured behind the dam, but, in so doing, radically disrupts the ecosystems established in flowing water. (See Ebenhack and Martínez 2013 for a discussion of "in-between" resources.)

One of the points that we want to make clear is that there is truly no free ride. Every human activity impacts our surroundings: our environment. The value of every energy source—and of every form in which we store it, concentrate it, or convert it to end use—must be viewed in the contexts of both the services it provides and of the costs of every step. Those costs include environmental and social impacts as well as direct economic costs.

Another essential point to bear in mind is that, although energy provides for many discretionary wants, it is also essential to our very survival. The value obtained from energy is not independent of the level of the needs it serves. Cooking and heating keep us alive. Lighting schools and hospitals is essential to supporting education and health care, respectively. Energy for industry enables the production of materials that allows for the goods and services of modern society.

Finally, in Affluent Societies, a good deal of the energy consumed goes to purely discretionary and luxury use: entertainment; exorbitant cars; and pleasure travel. There is a clearly declining progression of values added by each step in the preceding sequence. A society's place on the development ladder dictates the forms of energy required. Although some may argue wistfully about the advantages of pastoral societies, industrial productivity is mandatory to support modern population levels. Energy is an essential supporter of activities, but it must be transformed to the service required. The most basic uses of energy required simple transformations to heat and light. The transformations to modern services are more complex.

Let us put further perspective on the transformative nature of energy systems coupled to machinery. The world's population was on the order of about 1 million people at the end of the last ice age. By the time cities emerged, a few hundred years before this current era, the population had risen to over 200 million. At the beginning of the Industrial Revolution, the population had surpassed 300 million, which then skyrocketed to 1 billion people by 1900. Fossil fuels and electricity use were in full swing and today, a mere 115 years later, Earth holds a population of over 7 billion people.

Vaclav Smil, in his aforementioned text *Energy: A Beginner's Guide*, describes the transition from traditional to modern society quite eloquently. In a traditional society, he explains, humans drew their food, heat, and mechanical power from flowing water and wind, agricultural crops both for food and solid fuels, draft animals and human muscle for labor, and shrubs and young and mature trees for combustion. Aside from wind and water energy, which are real-time transformations of solar energy, all of these other resources took on the order of months or decades to harness. Things took time to do, and the relative scarcity of the resources allowed for controlled growth of industries and populations. It is fair to suggest that this societal mode is sustainable on the order of many millennia.

Modern society's goals are very similar to those of traditional society. They support the population's need for food, heat, and industry, but the energy systems currently available to us have dramatically changed both the scale and speed at which development can progress. Machines took their place in the textile industry and in processing food, especially milling grain. These changes increased production and the availability of goods, while relieving people of many tasks. The transition, while replacing some tedious human labor, certainly had its dark side.

The vast productivity of the emerging industrial centers created new kinds of jobs and drew workers into growing urban centers. For the most part, the jobs were menial and low paid, but a few people gained conspicuous success. Of course, the Industrial Revolution also came about in a society that was largely agrarian, but when very few of the people working the land owned any of it. Most of the populace were tenant farmers, who owned very little. Those families who did own small farms tended to have large families, providing many hands to work the farm, but also too many descendants for everyone to inherit enough land to make a living. People with little hope in subsistence farming flocked to the cities, in hopes of making their fortunes—dreams that seldom came true. The need for cheap, abundant labor was met by many thousands of people living in squalor, devoid of the subsistence agriculture they had left behind. An abject form of poverty became pervasive, with sprawling ghettos.

Once industrialization began, it demanded more and more energy inputs. Increased energy production inevitably brings with it increased environmental impacts. We cannot merely eliminate major energy sources. We need energy, and modern society requires a great deal of it. Current population levels likewise require a great deal of energy. Putting those factors together, tells us that we need to be able to continue to increase the value of energy produced globally until all people have enough energy to support reasonable qualities of life—and until population levels stabilize. Neither of those will happen quickly or easily. Energy production will have to increase for several decades, even as we find ways to conserve, that is, to produce more value with less energy.

All the while, we want, even need, to reduce the environmental and social costs of energy systems. New kinds of energy sources, without impact and without limit would be the ideal solution. It is tempting to look to renewables as the immaculate "white knights." Renewability does address depletion. It does not mean, though, that there are no limits, and there is nothing about the concept of renewable energy systems that mean that energy sources will be without environmental costs. We believe that this is one of the essential understandings to enable us to plan energy systems that enhance value.

Evaluating Our Choices

Valuing energy is not just a scientific or technical endeavor, as social, political, legal, and ethical attributes are also important. However, science and technology set constraints and provide a mechanism for quantifying value. The economy, governmental policy, and even human inventiveness must operate within the constraints of what is scientifically and technically possible.

Romer (1976), in his 1978 book, *Energy: An Introduction to Physics*, stated three facts about the use of energy that remain quite relevant today:

1. The energy we use has to come from somewhere
2. The energy we use has to go somewhere after we use it
3. Every energy conversion has side effects that may be undesirable and that may not have been anticipated

Let us look at these points a little more closely. Almost all of the energy that we convert and use on Earth can be directly (or indirectly) attributed to the sun. Direct solar radiation, wind, and biomass (including fossil fuels) are all solar energy and solar energy derivatives. Fossil fuels in particular are merely ancient stores of the chemical energy of plants, now being unlocked after millions of years. What distinguishes these resources is the manner in which we process them. Most biomass is burned directly, without processing—but very polluting. The fossil fuels have been processed by nature for millions of years and actually burn cleaner because of it. Major efforts have been underway to transform biomass into higher quality fuels. A significant amount of biomass is being converted to ethanol, primarily as a transportation system, which often incurs massive toll on land, water, and food systems, for very little net energy produced. Waste streams generally avoid these negative consequences.

Solar, wind, hydropower, and geothermal energies are generally converted to electricity for large-scale commercial applications. That has benefits to the consumer, providing very versatile energy forms, with minimal impacts at the point of end use, but do not be misled to believe that means that the systems are without impacts. The impacts are moved upstream to the manufacture of the devices and facilities, rather than at the end use. We ultimately want to—need to—optimize total systems to produce the most and highest quality energy, with the least overall costs.

As such, humanity is beginning to take on the gargantuan task of placing appropriate costs and benefits to maintaining our modern, energy-heavy society and using ever cleaner resources and technologies to do so. New alternatives to fossil fuels must be able to meet current energy use practices such as transporting people and goods, heating and cooling spaces, lighting, and cooking. The list goes on.

Fossil fuels, while perceived as dirty fuels, in actuality can be quite clean, especially when compared with the combustion of raw biomass. In poor regions across the globe, biomass consumption by far outpaces the consumption of any other energy source. This is by necessity, not by choice. Firewood dependence is dangerous, dirty, and limits development.

Within the subset of the fossil fuels, coal is surely the most polluting, and natural gas is by far the least polluting, which is why such an emphasis has been placed on exploiting the shales. The fossil fuels are often viewed collectively in very negative terms, which can obscure the values that they have brought society, and the costs that would have been associated with any of the other alternatives existing at the time. The transition from firewood to coal represented an advance. The transition to petroleum dominance was another gain. The growth of natural gas offers potential for even more improvement—gains that can proceed at a greater pace than seems likely for any of the nonthermal sources.

In trying to maximize the value added by our energy systems, we need to be able to evaluate and to compare the costs and benefits across the various components of the various energy systems. We need to develop viable, robust metrics to compare the costs and benefits across and within the various energy systems. There are a number of metrics that can be drawn upon. From life cycle assessment, to net energy value, to exergy, these metrics provide a basis for evaluating environmental impacts through the stages of production of any product good or service. They look at energy from the acquisition of energy, through its transport, storage, conversion, and end use. This is a significant development in evaluating resources at a systems level. However, they do not really address the benefits or values offered and none of them fully capture the range of costs and benefits. Particularly, we suggest that it is imperative to develop metrics that account for the values offered relative to a range of needs and how values change over time and development contexts.

We cannot expect to find the perfect resources, nor the perfect applications. We can seek to improve the balance of benefits relative to the costs. To do so, we must assess those costs and benefits as realistically, honestly, and fully as possible. In so doing, we must consider energy systems in their totality: the sources, the conversion, storage, and transport of energy. We must work to assess the value of current energy systems and what changes can enhance the total values, over what time frames.[5]

We contend that the economic, social, and environmental impact of human activity is most accurately measured by our consumption of energy, both in terms of type and quantity. Since shortly after World War II, we have seen an amazing transformation of the global economy. Public awareness of the impacts of human population growth and large-scale energy use, especially in the last 30 years, has caused attitudes to change over how to deal with environmental protection and social responsibility, placing value on our behaviors with respect to energy consumption and its impact on our surroundings.

Final Thoughts

As we have demonstrated in this chapter, increasing energy consumption has been at the foundation of human development and progress. Indeed, our current global, consumerist economy depends on massive quantities of primary energy to provide the liquid fuels, heat, and electricity needed

[5]There are also nonenergy by-products associated with most energy sources, but the by-products of petroleum production are particularly important and valuable. In modern societies, we are quite literally surrounded by petroleum products: clothing, cars, furniture, dishes, paints, lubricants, toys, food additives, and pharmaceuticals. Even if energy production from petroleum ended, its value as an organic chemical feedstock will probably continue for many years.

to fulfill our consumption choices, which now range from cooking and lighting, to heating buildings, to manufacturing our goods, to delivering services in person and online, to moving objects and people through air, land, and sea. The goods and services provided by energy have resulted in a higher quality of life. That much is certain. However, those very development benefits have, in turn, created more incentive to consume. This, in turn, has reinforced continued demand for more primary energy to support or improve on this high quality of life.

This pattern of increased consumption has put a clear strain on primary energy resources, and on the environment from which resources are extracted and to which wastes are returned. As long as our consumption was small in comparison to energy and environmental resources, consumption could grow with little attention paid to the costs and benefits of our choices. When local resource capacities were exceeded, some societies waned. Some societies utilized trade to expand their resource base. Some people innovated improvements to existing systems and found new energy resources to exploit. Today, our resource extraction and our impact are global. We must find means to gain the most benefit from our energy use that we can, while minimizing the negative impacts. We cannot simply say we will use less energy, as many people still need the benefits of modern energy. We cannot afford simply to stand in opposition to every negative impact of energy systems: for they all have those negatives. We must be deliberate in assessing our options and finding ways to optimize the benefits across cultures, geographical constraints, and time.

References

Blok, K. 2007. Introduction to Energy Analysis. Amsterdam: Techne Press.

Ebenhack, B. W., and Martínez, D. M. 2013. The Path to More Sustainable Energy Systems. New York: Momentum Press.

Khodeli, I. (Ed.). 2009. From Curse to Blessing?: Using Natural Resources to Fuel Sustainable Development. Hoboken: Wiley-Blackwell.

Lawrence Livermore National Laboratory (LLNL). 2014. "Energy flow charts." Accessed from: https://flowcharts.llnl.gov/

Maxwell School of Citizenship and Public Affairs. 2014. Available at: http://wilcoxen.maxwell.insightworks.com/pages/137.html

Meadows, D. H. 2008. Thinking in Systems: A Primer. White River Junction: Chelsea Green Publishing.

Perlin, J. 2005. A Forest Journey: The Story of Wood and Civilization. Woodstock: The Countryman Press.

Romer, R. H. 1976. Energy: An Introduction to Physics. San Francisco: WH Freeman.

Smil, V. 2012. Energy: A Beginner's Guide. London: Oneworld Publications.

Thomas, R. 2015. "Producer Gas Plants." Accessed from: http://www.academia.edu/194911/Producer_Gas_Plants

United Nations Development Program (UNDP). 2005. Energizing the Millennium Development Goals. New York: UNDP.

Yergin, D. 2011. The Prize: The Epic Quest for Oil, Money & Power. New York: Simon and Schuster.

CHAPTER 2

The Value of Energy Resources

Energy Resources

Some energy resources (i.e., the sources of energy supply) exist in great abundance, but are difficult to tap. Others are easy to tap, but not particularly efficient to convert to useful energy. Yet others are easy to tap, relatively efficient to convert, and currently abundant, but are not viable as long-term energy options. Factors such as these set a value to a particular resource that must be measured in a fashion relative to the service it provides within existing infrastructures, and the overall viability as a long-term and dependable, that is, sustainable, energy source.

For example, the well-known depletability of fossil fuels sets limits on their future value, as they cannot be replenished, while their present value is extremely high, due to the abundance of energy that can be readily accessed from them right now to serve existing infrastructures. On the other hand, solar energy's value right now is marginal due to the issues of intermittency and storage, and lack of an extensive infrastructure. Yet, the raw energetic content and long-term prospects of solar's utilization make it highly valuable going forward, probably most valuable when fossil resources begin to dwindle rapidly.

Indeed, myriad differences between the energy resources make it difficult to place set, comparable values on them. Estimating raw energetic content and abundance is relatively simple to do, but determining other essential factors such as availability and social value can be challenging. Thus, a key is to isolate core qualities that can then be used to assess an overall value for each energy resource.

Abundance and Availability

Energy is required in great quantities to meet the needs of burgeoning populations seeking to develop and to improve their qualities of life. Even the world's affluent seem constantly to want more goods and more services, using more energy. Without a doubt, industrial and postindustrial societies are built on the back of cheap, abundant energy. Thus, as energy becomes constrained or expensive, economic productivity inevitably is impaired and development stymied. Energy must be available in desired forms, with desired intensity, when and where needed.

As discussed in Chapter 1, the ability to tap the vast accumulation of millions of years of solar energy, stored in the form of fossil fuels, was critical for human development. Collectively, they offer vast storehouses of energy, refined and intensified through millions of years of bio- and geochemical processing. A barrel of crude oil, for example, contains enough energy to drive a typical automobile nearly 1000 miles, even before it is further refined into gasoline or diesel constituents. Indeed, many oil wells start out flowing thousands of barrels per day; enough energy for millions of automotive transportation miles, from a single well.

Thus, the concept of a useful energy source is based on abundant natural resources from which humans can draw to gain useful energy for personal benefit. Abundance in this instance, then, refers to how much energy we can access and, ultimately, how much useful work we can obtain from them. Availability, on the other hand, is a function of context; where and when the energy exists and how it can be tapped. Energy may exist in abundance and yet not be practical to utilize. That means that the true abundance is a function of how much more usable energy is obtained from a resource than is required to obtain it.

Resource Base

The abundance of stock-limited resources can be characterized by their resource bases, which describes the amount of each source believed to exist within a given geographical context. This term is generally used for the energy resources that have reserves, especially the fossil fuels and uranium. But we can also talk about the resource base of the flow-limited resources, albeit more vaguely. The resource base for the solar resource, for example, can be viewed as the total average flux of solar energy striking Earth's surface, whereas the resource base for wind energy can be viewed as the total movement of air in the atmosphere. Geothermal energy is more problematic in terms of quantifying its resource base. Should it be the entire heat of Earth's interior, or the heat to certain exploitable depth limits, or just the heat contained in unusually hot shallow reservoirs? Indeed, it cannot be all of the heat, because only temperature differences can be tapped to produce useful energy.

Admittedly, the resource base is a very rough estimate and impossible to know precisely for most energy sources, but it is a real quantity. Two key factors that affect the estimate for finite resources, in particular, include production and exploration. That is, the actual, remaining resource base only changes as it is reduced by production, since it includes all of the resource that exists at any given time, while exploration can expand our knowledge of where resources exist and improve our estimate of the resource base. As more of the actual resource base becomes available for human use, estimates of its size commonly grow. For the fossil and nuclear fuels, the resource base is equal to the total stock of the resource accumulated inside Earth. In a human time frame, it would be impossible to exploit more than the accumulated stocks of any of these resources. Moreover, some oil and gas will always be left behind in the reservoir rocks, and some will undoubtedly go forever undiscovered. Likewise, coal inevitably will be left behind in every coal mine.

On the other hand, the resource bases for resources such as solar power and wind power, which are not depleted at all by human use, are best considered as the total flux, or flow, of that resource on an annual basis. However, these flow-limited resources further invoke the issue of practical availability. If we were somehow to tap all of the energy from the sun, it would deprive plants of sunlight for photosynthesis and deprive us of the light that warms our bodies. It would also stop the winds, which are primarily a result of the uneven warming of Earth's surface, from the sun. Thus, we may be able to refer to the entirety of solar radiation, or wind, or Earth heat that exceeds ambient surface temperatures as comprising their resource bases, but we must bear in mind that a vast portion of the resource base is irrevocably unavailable.

Reserves

Reserves are typically estimated for each oil, gas, or coal field, or uranium prospect. Global reserves would be a summation of all of these prospects, and needs constant revision. By convention, production is subtracted from the yearly estimate, while new discoveries or enhanced recovery potentials are added to it. For oil and gas in particular, field reserves are reevaluated, based on the new information gleaned from production data, even in well-established fields. When new kinds of reservoirs are found to be productive—the shale plays of the 21st century being a notable example—there is a great deal even for experienced professionals to learn about how the fluids exist within the pore spaces and how they flow.

Indeed, reserves estimates are probably the number that can be known with some confidence, especially the aforementioned proved, recoverable reserves. First, note that this definition absolutely precludes any undiscovered petroleum. (If it has not been found, it certainly cannot have established production.) So, global, or national, or corporate reserves are constantly being added to by new discoveries. The caveat, of course, is that every barrel produced from a field must be subtracted from the reserves.

Higher prices inevitably lead to more aggressive exploration, which adds new reserves. Higher prices also support more exotic, and, thus, more expensive recovery technologies. The typical oil field, under conditions at the beginning of the 21st century, will leave behind anywhere from two-fifths to two-thirds of the original oil in place at the end of the field's economic life. That means that reserves are always much smaller than the amount of oil known to exist. It also means that a great deal of oil can be added to the future reserves picture as new technologies succeed in displacing more of the original oil from the tiny rock pores in which it resides. Reserves are dynamic numbers, being added to by technology and discoveries and drawn down by production.

Of course, the flux-based resources have no reserves to speak of, although hydropower, a notable "in-between" resource, does have a quantifiable, stored energy potential, especially in the case of large, dam-based hydropower installations. Biomass, too, has the potential to be classified as a resource with known reserves (i.e., forests, arable lands, etc.), which can be depleted if the rate of extraction is greater than the rate of renewal via photosynthesis. However, solar and wind resources, the resources with the greatest resource bases, have no reserves, which can, at least at first blush, make them problematic as baseline energy sources.

Delineating Definitions

To distinguish the differences between a resource base, a reserve, and production, let us consider some energy source with a resource base of 100 arbitrary units. Within the 100 units that truly exist in the world, there are two parts: (1) one that is known or believed to exist and (2) the other that is not recognized. (It is actually possible that optimistic researchers will believe more of something exists than truly does, which would mean that there is a portion that does not really exist. Engineers and scientists are trained to think conservatively, so it has generally been the reality that resource base estimates grow as we learn more about the occurrence of each energy source.) Within the portion that represents the estimated initial resource base, there is cumulative production and remaining estimated resource base. Since that which has already been produced is no longer available, it is not part of the remaining resource base. Within the remaining estimated resource base, then, is proved, recoverable reserve (see Figure 2.1).

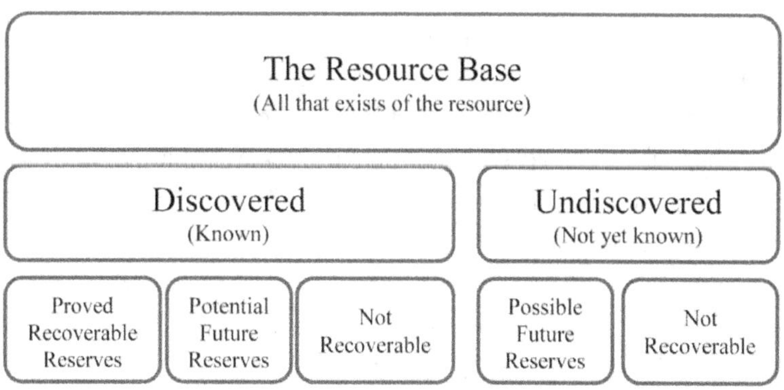

Figure 2.1 Visualization of a resource and its reserves.

To better conceptualize discoveries adding to, and production being taken out of the resource base, think of playing blackjack in a casino that is using many decks of cards at once. If you know there are four decks, then you know that you start with 16 aces. With each successive ace that is played, it gets harder (less likely) for you to be dealt one of the dwindling aces that remain. As more and more of the true resource base for a given finite resource is identified and brought into production, it gets harder to find the remaining fields, since there are fewer of them. As of this writing, there are still many aces in the deck of oil and gas exploration and discoveries continue to abound. Someday, that will not be so and reserve additions will not keep pace with production. At that point, the world will be entering the peak production period, and the currently unseen limits will be felt.

Availability in Time

Energy demand has been increasing throughout human history and will undoubtedly continue to increase until there is a stable global population, and all people have reasonable access to relatively high value energy systems. More energy is needed for now, while it is imperative to understand the limitations on energy resources we tap. On one hand, the production from stock-limited resources will necessarily reach a peak value and begin to decline, reducing availability over time. On another hand, flow-limited resources are not depleted by human use, but their availability is limited by the daily flux of the resource, which varies from one site to another. On yet a third hand, the technologies to bring new resource systems into production take time to develop, which also affects availability.

The resources on which we now most rely are stock- or reserve-based. The stocks that we can draw on as needed offer great value, but our use does deplete them. Some of the possible resource systems of the future are not technologically available yet and we need to consider the time to maturation for their requisite technical development. Some resource systems draw on energy flows that our use does not deplete. They face two kinds of time constraints: the time to scale up production technologies and the limits on availability due to the variability of their flow rates.

The Effects of Energy Shortfalls on Value Provided

Much of the conversation about sustainability regarding fossil fuels is related to the ultimate abundance of the fuel and really stems from the oil crises of the 1970s (see Horowitz 2005). People feared that the end of the petroleum era was in sight. This notion was supported by highly publicized reports that used the "production to reserve ratio" metric as an indicator of when the world would run out of petroleum. If global petroleum reserves were 635 billion barrels, as they were in 1973, and global production was 21 billion barrels, as it was in 1973 as well, then it would take 33 years to deplete those reserves. It seems simple enough, right? But 40 or so years later, rather than being 8 years past the end of oil production, the world is producing some 40 percent more oil than it was at the time of the great embargo.

How can that be? Watkins (2006) suggests that it is a failure to understand oil and gas as economic commodities, provided by the marketplace. We suggest it is more to the point to focus on the true meaning of reserves. As we have seen, the term reserves does not refer to the total amount of something that exists, but rather the amount that is currently known to exist and believed to be recoverable under existing economic and technological constraints. The reserves-to-production ratio of a reserves-based energy source is only significant if the ratio decreases. Even then, we suggest it is more significant to consider the changes in surplus production capacity, a term that seems to be hardly ever used.

Global surplus production capacity is the difference between global demand and maximum global production, which is a summation of the maximum production rates of all the wells in the world. An individual well's maximum production is a function of the conditions of the geologic reservoir in which the oil and gas are constrained. In essence, it refers to how much of a resource can be produced at any given time, in excess of the amount the marketplace demands at that time.

Our current economic system needs surplus production capacity of oil and gas. Without it, any supply disruption—having to shut wells in for workover maintenance, or pipeline damage, or an embargo, or a host of other problems—would create shortages. Every shortage would drive prices up suddenly. Then, prices would tumble for a while as production was restored or new production became available. In this frenzy of rising and falling prices, some wells would, inevitably, not be worth producing during the low-price cycles; it would cost more to operate the well than the value of their production. Hence, operators would shut such wells in during low-demand cycles and bring them back on line when shortages drove prices higher. The balance would re-assert itself, one way or another. However, if companies made no effort to watch demand cycles and produce accordingly, consumers would be faced with highly uncertain prices, not only for gasoline, but for other energy services. (Natural gas prices might triple or quadruple in cold winters and drop very low when people needed little gas in the summer.)

Oil and Gas Limits

The total resource base of petroleum is, of course, unknown and unknowable. However, informed estimates put it is somewhere on the order of 20–30 trillion barrels. That is about 10 times the proved global reserves. (As of the close of the 21st century's first decade, the global annual production was about 30 billion barrels.) At that rate, the global petroleum resource base would be depleted in about 1000 years. That's a long shot from the 20 years suggested by the estimate of the reserves-to-production ratio. The difference helps to demonstrate the enormous difference in these terms. It is also a function of the fact that we will never be able to find, let alone produce, the entire resource base. We will certainly not be producing crude oil at current rates 1000 years from now.

Production will continue to grow as long as it is able. That is, as long as new discoveries and technologic improvements can more than offset the oil produced from existing reserves. For more than 150 years, reserves have grown, because depletion was more than offset by additions from discoveries and new technologies. The recent technologic developments to produce oil and gas commercially from the extremely low-permeability shale rocks are leading to huge reserves additions, which will extend production remarkably. However, when additions can finally no longer fully compensate for production, then global petroleum production will peak and begin the long, gradual decline (see Schölnberger 2006).

Peak Oil

Growth in petroleum production has been driven by two things: a nearly insatiable thirst for energy to fuel economic development (particularly in transport) and an unprecedented capacity for production. Indeed, nothing before was ever able to produce in the quantities that petroleum has been able to achieve. And the inevitable decline of petroleum will be a function of increasing difficulty in finding prolific new reservoirs, and of the natural decline rates as reserves are depleted from the world's petroleum resources.

When a global production peak approaches and petroleum is challenged to support growing demand, crude oil prices will rise dramatically, no doubt over a short time period. If history is any indicator, this will encourage tremendous investments in producing more. Exploration will increase and more technology will be applied to increase the production of the fuel in short supply. This will tend to slow the decline. Indeed, it may temporarily halt or even reverse the decline, as seems to be the case now. However, at some point, the finite nature of the resource will dictate that reserve additions begin slipping below production rates. We suggest that until flow-limited resources are able to take up that deficit, petroleum will continue to be relied upon heavily.

When the maximum production rates of all the wells in the world no longer exceed demand, the surplus production capacity will be exhausted and the world will be entering the period of "peak oil." Prices will rise, incentivizing increased investments in exploration and improved recovery technologies. These will add more productive capacity, but depletion will continue to offset these gains. We envision an unsteady plateau as higher prices add more production, but depletion continues to take a toll. It will be like blood transfusions for a leukemia patient. Each transfusion, in the form of a new field discovery, will offer some relief, but each cycle will tend to offer less relief for a shorter time, until the final decline begins.

Remember, oil and gas reservoirs are depleted by production. As fluids flow from the surrounding reservoir rock into the wellbore, the pressure declines in the reservoir surrounding the well and the flow rate begins to decline. Although the process of generating fossil fuels is, in fact, ongoing, its rate of renewal is orders of magnitude less than the rate at which humanity extracts them. We have discussed the fact that the depletion is offset by reserve additions, which come from new discoveries and technologic improvements. For individual oil and gas wells, though, the production rate is a direct function of the pressure difference within the reservoir. The pressure declines as more fluid is extracted from the reservoir rocks, thus the maximum possible production rate for the well declines.

The understanding of peak oil has been drastically changed by the rise of nonconventional oil and gas resources. "Conventional oil" is a frequently used term that may become obsolete rather soon, but not because oil will be obsolete. "Conventional" originally referred to liquid hydrocarbons that had accumulated in the void spaces of naturally porous and permeable rock, essentially sandstones and carbonate rocks. When a well is drilled into a conventional oil reservoir, oil can flow naturally out of the pores into the lower pressure wellbore.

Unconventional oil includes deposits with extremely low natural permeability and very heavy organic materials that are too thick to flow naturally (tarlike or even solid). There is a continuous spectrum of rock types and of organic material. Rocks may range from extremely impermeable shales (with permeabilities measured in a few billionths of the permeability unit, darcy) through moderately permeable rocks, and even to the best reservoir rocks (having natural permeabilities of several darcies.)

The wells with long laterals expose much more of the reservoir. Since these wells are drilled in the shales, which are quite impermeable, no real flow to the wellbore is possible without fracturing the rocks. It is a combination of drilling long horizontal legs through the shale and then fracturing those long intervals that allows recovery from the extremely low permeability shales. These combined technologies began in the early 21st century to move a large segment of these oil and gas resource bases into the reserves category. The combination of being able to steer long horizontal legs on wells, of more than a mile, along with extensive, multistage fracturing techniques can create significant permeability along extensive fracture planes.

The shale "plays" are now adding extraordinary resource value to oil and, especially, to natural gas. The total resource base has yet to be well-defined, but in the United States, where the plays are especially mature, it appears that 40% of current reserves are already from the shales (EIA 2014). This is somewhat less impressive than it could be, since United States peak conventional production has passed. Understanding of the shale resources is growing, and, historically, the growing understanding of a resource leads to increases in estimates of its abundance. We suggest that it may be reasonable to think that the global natural gas ultimate reserves will increase by about 60% due to these plays.

In spite of some authors arguing to the contrary (see Deffeyes 2010), a great deal of petroleum remains to be discovered. Enhanced recovery techniques also stand to, well, enhance recovery. With that in mind, we can envision using an estimate of ultimate recoverable global reserves of 6 trillion barrels. That estimate is three times what most of the popular peak oil theorists have suggested, but it is in line with estimates of years past (Grossling 1977). After all, most of the African continent remains unexplored, as do large parts of the Arctic, South America, central Asia, and deep waters. It seemed reasonable that those vast tracts would yield more oil and gas. As of this writing, with shale plays revolutionizing the oil and gas industry, that 6 trillion barrel estimate seems less and less conservative.

Peak Gas

Natural gas is closely associated with oil; geologically, they may occur together or separately. And like oil, natural gas will also reach a peak in production and begin to decline. The natural gas peak will probably follow the oil peak by a few decades, due to a number of factors, but mostly because utilization and development of the resource on a wide scale began later. For example, Vaclav Smil points out in his book, *Energy, Food, Environment*, that natural gas was long ignored, in favor of the readily contained and transported liquid fuels (Smil 1987). Innovations in gaseous pipeline technology, though, proved extremely effective for moving natural gas within continental boundaries.

A shocking amount of resource value has been squandered over the first century of the petroleum industry as gas that was produced in association with oil was flared (burned and released directly into the air). Tragically, the practice of flaring continues, especially in very poor countries where markets are not adequately developed to take advantage of the resource.

Fortunately, the value of gas is being increasingly recognized and it has the potential to serve many of petroleum's traditional roles. Gas is already the preferred fuel for space heating and for cooking in much of the developed world. It has serious potential to be produced and used to support local needs in Developing Countries, in order to displace firewood use.

As gas has come into greater use, its virtues have become apparent enough so that it became worth building pipelines and related natural gas transmission infrastructure throughout much of the affluent world. The demand for gas grew especially fast for domestic use. It is far more effective to utilize gas for cooking and for space heating than to utilize coal or firewood. No more getting up at night to stoke the furnace. No more building a fire and waiting for it to sufficiently cook. With the appropriate infrastructure, gas flows readily to the point of use, as needed. And it burns much more cleanly than the solid fuels.

By the second half of the 20th century, gas began to be used in more industrial applications, including generating electricity. But the oil shock of the 1970s Oil Embargo era raised the specter of impending energy shortages—an Energy Crisis. Natural gas use in generating electricity was curtailed, to save it for distributed applications, especially domestic use, where scrubbers and pollution remediation devices would be prohibitively expensive.

In 2014, the United States began promulgating regulations to reduce carbon emissions, which are likely to push for increased use of gas in generating electricity again (EPA 2014). There is a conundrum here; the choice of whether to displace coal with gas for generating electricity depends on the relative values placed on climate change versus the future availability of a clean efficient resource. Natural gas is certainly desirable for domestic applications of cooking and space heating. All of the gas burned to generate electricity becomes unavailable for future use. On the other hand, no alternative system other than nuclear fission has the production base to be able to replace very much coal. If our values do not favor rapid increases in nuclear fission, then gas substitution is one of the most effective ways to reduce greenhouse gas emissions in the short term. And because of this, natural gas, like oil, will reach a global production peak and begin a long, inevitable decline (see Maggio and Cacciola 2012).

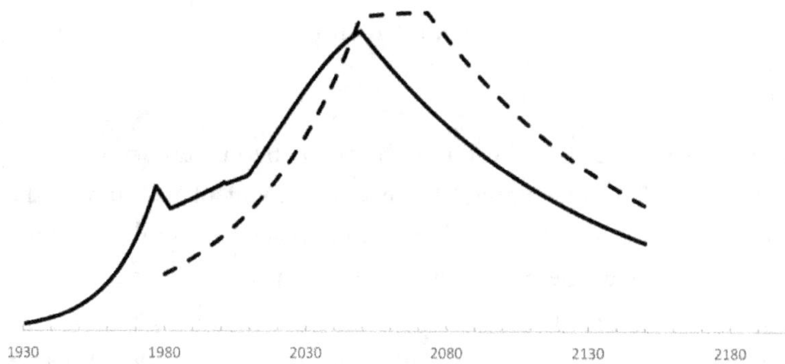

Figure 2.2 Possible oil (solid) and natural gas (dashed) peaks. Historical data from US EIA DOE International Energy Statistics.

Figure 2.2 illustrates possible peak scenarios for oil and gas together. It makes sense that we should look at them together, as they occur that way in nature. They are ultimately quite substitutable. Since natural gas has not been as highly valued as oil, its peak will lag behind oil's by a few decades. Of course, the curves above are built on several assumptions about the relative sizes of the two peaks as well as about how many decades gas lags behind oil. The curves are not presented to suggest that we have the absolute answers, just to suggest a character of what to expect.

The gas peak lagging behind oils offers some good news for the future. Continued growth in gas production will help to soften the shortage—probably much more than will the growth in solar and wind production by the middle of the century. The shale plays have offered significant recent growth, especially in natural gas production. That growth in gas production (alongside a global recession) has driven gas prices down in the early 21st century. With the lower prices, gas is being viewed as an alternative to coal for power generation and, increasingly, as a transportation fuel.

We stress that oil and gas production estimates and forecasts are probably best viewed in combination. It might seem logical, then to treat oil, gas, and coal together in a composite fossil fuel peak. However, the limits to coal use are likely to be a function of its social and environmental costs, rather than its physical limits. Furthermore, oil and gas occur together. Therefore, as we evaluate the transition and limits to growth of alternative energy systems, we will compare based on a composite oil and gas peak.

Other Limits

Nuclear Resource Limits

Nuclear fuel material is also directly depleted by our extraction. Some optimists are fond of pointing to the total amount of uranium in the world, as evidence that our use of nuclear fuels causes very little depletion, but the richness of the reserves is an important factor. Where uranium occurs in thick veins, it is reasonably easy to extract. However, vast portions of the uranium resource base are dissolved at very low concentrations in the waters of the oceans. It may be technically, and even economically, feasible to extract uranium from solution for a while. However, once half of the uranium has been extracted, what remains will exist at half the initially very small concentration and, thus, be at least twice as hard to extract. Depletion of high-quality accumulations will make future nuclear fission much more costly, as no uranium is generated within Earth.

Without doubt, uranium's resource value would increase dramatically if our conversion technologies improve so as not only to use very rare fissile fraction of uranium (U-235), but also expand to using the fertile fraction of U-238, which is more than 100 times more abundant. It remains finite, though, and will ultimately be depleted by excessive use. This would constitute a technologic innovation to expand the technically feasible resources beyond what is currently conventional, much as the shale plays have done with oil and gas.

Flow-Limited Resources

Solar and wind resources have no stocks on which to draw, or to deplete. No matter how much energy we garner from these resources, our use will not diminish their continued flow. The sun apparently continues to shine and the wind apparently continues to blow day in and day out. (Thus, the usage of the term "renewable" to describe this fact.) Of course, everything in the universe is finite, including our sun. It has an enormous stock of hydrogen. As the sun's huge gravitational force holds the hydrogen ions so tightly together that they undergo fusion, the sun will effectively burn itself out, albeit in a few billion years. Solar (and wind) energy will disappear, but Earth itself will probably be engulfed by the sun as it expands in its later days.

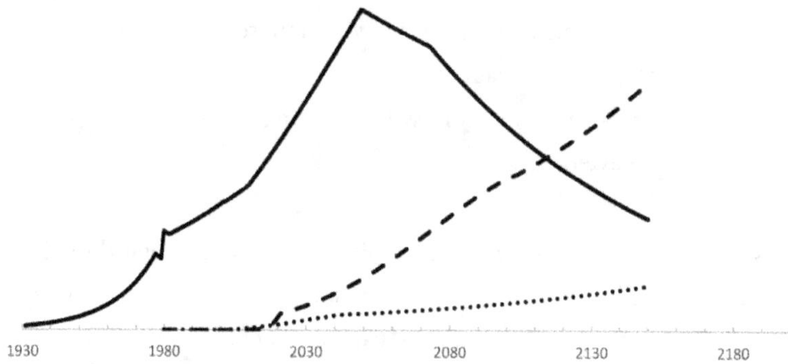

***Figure 2.3** Composite peaks showing possible growth trends for wind and solar in comparison to a combined oil and gas peak. The dashed line represents a solar model for 30 percent growth to 2020 and then Gaussian to 2100. The solid line represents a combined oil and gas peak. The dotted line represents a wind model for 26 percent growth per year, then Gaussian growth until 2050 at 30 Quads and 1 percent per year afterwards. (Data from US DOE EIA International Energy Statistics.)*

Consider Figure 2.3, a composite of oil and gas peaks, compared with wind and solar energy production. The figure is based on several realities:

- Dramatic growth of solar power has a very short track record. How long can this unprecedented growth rate be expected to continue?
- Wind showed a short history of extreme growth (basically two 5-year spurts of growth rates above 25% per year) but may be showing saturation behavior already.
    - The last few years have showed considerably slowed growth in wind power production.
    - Wind power has saturated a number of the prime locations where high average wind speeds are in proximity to consumption centers.
    - Wind power reached scales at which some of its negative externalities have become more apparent.
- The rate of solar growth can be expected to exhibit some form of saturation behavior.
    - Locations will not saturate readily, since sunny locations are popular.

- Larger scale manufacturing will strain some of the needed materials, especially rare earths.
- Large-scale operations will also bring attention to negative externalities.

Wind and solar will grow, but they will still be very challenged to offset the declines in oil and gas in time to avert shortages. Their peaks, though, will not face the inevitable decline that characterizes fossil fuel peaks. Rather the peak for solar and wind will be decreased growth, due to market saturation and to waning enthusiasm as negative externalities are realized when scales of production have increased by more than a factor of 10.

Technical Viability

Technology is required to tap all kinds of energy and there are limits imposed by technologies required. One of the reasons that very little oil or natural gas is produced for local use in developing countries is that very specialized technologies are employed in finding, drilling for, completing, and producing oil and natural gas.

Technologic development can and does make more of the resource bases available. Technology is critical to finding underground fossil, nuclear, and geothermal resources. It is vital to acquiring all kinds of energy for use. The rapidly emerging production of oil and gas from tight shales is a dramatic example of how the application of technology greatly increased the availability of a resource. Technological development in the solar and wind energy industries is also a major factor in making more of the resource accessible. Efficiency conversions and load factors have dramatically impacted the viability of wind energy, particularly in recent years.

Energy Quality

Not only does quantity matter, but quality as well. The value of any resource systems is also a function of its intrinsic qualities, which determine how it can be used. Of particular interest are the energy density, the power density, and the related dispatchability of the resource.

Energy Density

"Energy density," also known as specific energy, relates primarily to intensity. It can describe either how much energy is carried per unit mass or per unit volume, typically expressed in units of mega joules per kilogram (MJ/kg). It basically asks the question: How much mass (or volume) do you need to provide a given amount of energy? This is an important characteristic for the ability to move energy to where it is needed.

Consider raw biomass that has not been refined by natural or human processes. Firewood can be burned directly, but in biomass-dependent regions charcoal is preferred. Charcoal is woody material that has been processed by heat to drive out most of the water. It burns hotter and cleaner, with a much higher energy content per pound. That makes it easier to transport. The fact that it readily breaks up into small chunks also makes it easier to select the appropriate amount for a given task. People pay a premium for the increased energy intensity and combustion performance for charcoal rather than firewood. Wood must still be harvested in a laborious process, but the resulting charcoal product can be moved readily to urban markets, where it is a prized commodity.

Note that "coal" is part of the word "charcoal." It is not a coincidence. Coal has also undergone processing, which drives off water, increasing the energy density of the fuel. It delivers more energy to the task than unprocessed biomass. Crude oil is even more processed by nature, and natural gas the most. Of naturally occurring combustion fuels, natural gas has the greatest energy content by mass (and the cleanest burn). It is so light, though, that it must be compressed under high pressure to hold nearly as much energy per unit volume as the liquid or solid fuels.

The only natural energy source with greater intensity than natural gas is nuclear fuel. While uranium is rare, and the naturally fissile isotope (U-235) represents less than 1% of uranium, it takes very little fuel to yield a great deal of energy. The limits on nuclear fuel use currently have less to do with the abundance of the fuel, and more to do with social factors stemming from anxieties about waste, accidents, or misuse.

Another fuel worthy of consideration is hydrogen gas. Hydrogen has the highest theoretical chemical energy density of any nonnuclear fuel (see Table 2.1) and is the most abundant element in the universe. It burns

Table 2.1 Specific energy of various fuels.

Fuel	Specific Energy (MJ/kg)
Uranium (0.2% in ore)	1400
Hydrogen (derived)	142
Natural gas	54
Diesel, gasoline	46
Anthracite coal	33
Charcoal (derived)	30
Wood	18

even cleaner than natural gas. However, hydrogen is not naturally available, but exists in the chemical bonds of countless compounds. Thus, it must be unstuck before being useful, which reduces the net energy density of the fuel.

Power Density

"Power" is the flow of energy per unit time; thus, the term power density refers to how much energy can be drawn from a given volume or mass per unit time. Within the context of energy resources, power density can be viewed as the rate of energy production per unit of earth area, typically in units of watts per square meter (W/m^2).

According to Smil (2006), fossil fuels, which have had the benefit of millions of years of natural processing and refining of biotic material, can be extracted with massive energy flows for small areas supplied. A coal mine, for example, can supply as much as 100 W/m^2, while an oil field can supply as much as 1000 W/m^2. This is a massive improvement over "real-time" biotic energy. Forest-harvested wood, for example, has a power density closer to 0.2 W/m^2, while the refined wood product charcoal has a power density of 0.04 W/m^2.

How do these compare with solar power density? We know that the power density of solar radiation when it reaches the upper atmosphere is 1370 W/m^2. However, since earth is a sphere, this number is reduced by one-fourth to 342 W/m^2. Moreover, reflection from clouds, ice, and liquid surfaces, as well as back scattering, and absorption by the atmosphere reduces the solar radiation "constant" to 170 W/m^2. If a solar collector operates at 15% efficiency, then this power density number would further reduce to about 25 W/m^2. While better than biomass, this number is much smaller than fossil energy, and is substantially reduced from its original value, speaking to the fairly diffuse nature of solar and solar-derived resources located on or near the earth's surface. (Geothermal energy located inside earth is equally diffuse, for similar reasons.)

Dispatchability and Serviceability

"Dispatchability" refers to the ability to deliver energy where, when, and how it is desired. Some energy needs are sufficiently discretionary to be set aside for times when it is most readily available, but many are not. One of the earliest benefits that external energy offered humanity was the ability to work into the night. Transportation inevitably requires mobility of the energy itself. Energy that can be stored is naturally more dispatchable than energy that cannot be stored. This is one of the value factors that favors the fossil fuels, which have been concentrated and stored in the Earth's crust for millions of years.

"Serviceability" has some correspondence to dispatchability, but it is more related to the ease of controlling its use. Is it convenient to use? Energy carriers, such as electricity, offer tremendous convenience value. Some sources, though, have more intrinsic convenience for direct use or for conversion to more serviceable carriers.

Natural gas is innately serviceable and convenient. It can be burned directly, with little to no processing, while producing intense, clean heat. It can be used at a very wide range of scales, from a tiny pilot flame, to an industrial boiler. Many people in the Affluent World enjoy the convenience of cooking with gas. Simply turning a dial to open or restrict a valve allows for anything from cooking over intense heat to warming over a tiny flame. We seldom think about the fact that it burns so cleanly that we don't even need to vent the combustion gases.

We can also use natural gas in large-scale industry, including generating electricity. Indeed in the early 21st century, there is some movement in the United States to undertake a significant shift from coal-fired power plants to gas-powered plants. This would make a substantial contribution to reducing the carbon footprint in the United States. When the intrinsic convenience of natural gas is taken together with newer gas turbine combustion technologies, it is possible for natural gas to "follow load," that is, for turbines to adjust their operation to respond to the ever-changing demand on the electric power grid (Masters 2013).

The question remains as to whether the best value is gained by using an intrinsically convenient source to generate somewhat more convenient electricity, rather than to save it for use in domestic heating applications, where coal and raw biomass would serve very poorly. It is especially important to raise this question with the increasingly more realistic prospect of using renewable energy sources in dispatchable power plants used to produce (relatively) emission-free electricity. Wind energy is extremely cheap these days, and installed solar costs have plummeted in just a few short years. The prospect of using fuel cell technology in dispatchable, stationary applications is also much more feasible.

Resources versus Income

A country may use its natural resources for internal development needs or for export earnings. Money earned from exports can be used to purchase goods and services, or so goes the thinking. In spite of the reality of development that actually happened in Affluent Nations, there is a common belief that resources are primarily of value in the Developing World to exchange for cash earnings, as a means to purchase what they need. This follows theories of competitive advantage.

The idea is that every country should focus on producing whatever they can produce more cheaply than others, then sell their product and use the earnings to purchase whatever they need, money being an infinitely versatile medium. In their seminal work, *For the Common Good,* John Cobb and Herman Daly argue that the concept of economic advantage may not be such a straightforward benefit (Daly 1994). The terms of trade really do not favor the exporters of raw materials. Many countries export raw materials only to spend a great deal more money to import finished products. This can include energy resources.

Indeed, in regard to energy, many of the world's least affluent nations that export energy, commonly in the form of crude oil, re-import refined products. Energy is an essential commodity for human survival and development; therefore, it begs questions about the relative value of money as opposed to the value of having locally produced energy available to meet local needs. For nonessential goods, export for currency to be able to purchase needed materials may add real value. It seems quite reasonable to question the concept of diverting essential resources to export, though.

Inevitably, there is profit-taking in every step of economic transaction, extracting some value. Consider food goods. The international community, through entities such as the World Bank, have, for decades, strongly urged Developing Countries to transform their predominantly subsistence agriculture into cash cropping for export. Chronic malnutrition is rampant in the countries following this advice. Some scholars of development have speculated about ulterior motives behind such guidance, to foster neo-colonial trading partnerships, in which the Developing World produces raw materials to ship to the Affluent World and to purchase finished goods back from the Affluent World (see Humphreys et al. 2007).

Subsistence agriculture is transformed into cash cropping. Food for survival is transformed into cash. Which is more useful? Crude oil is shipped out of countries and transformed into gasoline and diesel and a host of other products, which are essential to development activities and, thus, must be re-imported at higher prices. This has resulted in large shares of foreign exchange deficits for the importation of refined energy products in some Developing Countries.

The merry-go-round of exporting resources to earn money in order to be able to purchase imported resources may ultimately lead to nations holding on to more of their own resources for their own use. Critical resources should be the priorities unto themselves. International pressures urge export for the cash earnings, but we have seen very little progress as a result of this practice in most countries through the years. We suggest that development has been impeded by the folly of not differentiating between discretionary and essential resources. It is one thing to earn cash exporting gold and diamonds, even many mineral resources. It is altogether another thing to trade that which you need to survive for cash. Exporting energy has created a great deal of wealth, but it has seldom succeeded in spawning successful, nationwide development.

We suggest that resources are intrinsically more valuable than the cash paid for them. If that were not true, why would affluent nations be willing to pay for them? But beyond that, it is troublesome to assume that the marketplace operates perfectly in allocating value. The people of affluent nations strongly desire cheap and abundant critical resources. The organization of petroleum-exporting countries, OPEC, was originally founded to provide some power balance to Western interests, which tended to act as a monopsony, maintaining low prices on energy to support economic growth in the industrial consumer nations (Yergin 2011). OPEC does, indeed seek to control production, to maintain stable prices that can return real wealth to the countries who are exporting their energy wealth. This effort also serves to stabilize the market, mitigating chaotic price fluctuations that have plagued the petroleum industry since its infancy. But, have the efforts of OPEC ministers led to significant, sustainable development in their own countries?

In part, the earnings are often not broadly distributed throughout the population. Most of the oil and gas operations occur on government lands or seas. Thus, the vast wealth goes to the government officials. As anywhere in the world, fabulous wealth is a great draw to greedy, and dishonest, people. In some cases, this vast flow of wealth through a few hands leads to corruption. Even if it doesn't, it is likely to lead to some conflict and resentment. Even within the United States, some who own farms in the regions of the recent shale drilling boom lease their mineral rights for far more than the farm would have sold for. They observe that the landowners, with mineral rights, tended to be very pleased with the companies they were getting nice profits from the shale boom. However, neighbors who did not see the wealth, but only the traffic congestion, the noise, etc., while seeing their neighbors getting wealth that they could not, tend to resent the intrusion of the oil companies.

We suspect that the phenomenon is very similar in Developing Countries, where the bulk of the population lives in abject poverty, and then is confronted with disruption to their lives and land, while watching others become tremendously rich. Great imbalance is unsustainable. Africa in particular is rife with examples where only few benefit from oil revenues. It would benefit to change this mentality. Perhaps it is finally being recognized, as the president of Uganda recently suggested that they were "going to use the oil revenues to fast track the country's transport and energy infrastructure development, a key factor to unlocking the country's economic potential" (The New Times 2013). Doing this could provide the lasting benefit of developing energy resources for internal use.

Perhaps one of the most promising things about renewable energy systems is that they are more likely not to be exported. Most fossil energy production in the Developing World has been targeted as a means to generate export earnings, yet in many cases, these earnings have failed to stimulate development. Energy systems that rely on tapping the natural flow of energy from the sun or wind cannot yet be readily stored or transported. Thus, the energy generated could offer much greater value for local use and there is less pressure to export the product. (However, even this notion is being tested in northern Africa, where projects are being developed to export solar electricity to European markets.)

Energy is an absolute requirement for any successful development activity. Therefore, we suggest that using locally produced energy locally will prove to have greater sustainability value. The local use will have higher efficiency. The energy itself is more difficult to steal than money is. The benefits are more likely to accrue to the people in the region where the energy is being produced. All of these factors suggest that local power generation from flow-limited energy systems may have great value in support of development.

The potential advantages of using locally produced energy to support local development are plentiful, but go beyond that to advance the concept that decentralized energy systems are advantageous in general. Indeed, the pursuit of decentralizing energy to meet development and environmental goals is one of active pursuit across the globe. The World Alliance for Decentralized Energy (WADE 2014), for example, is an organization that "...works to accelerate the worldwide development of high efficiency cogeneration, onsite power and decentralized renewable energy systems that deliver substantial economic and environmental benefits…". This description matches well with our notions on what decentralization could achieve. Decentralized or distributed systems are less subject to large-scale disruption, largely immune to embargos, and have fewer losses in transmission. When energy is produced where it is to be consumed, it tempers prospects for export. This helps to ensure that it is used locally. However, local production can make local consumers more vulnerable to supply disruptions.

Tapping High Value Alternative Resources

Some of what we call renewable resources actually are reserve-based. This gives them an advantage in terms of being able to tap the reserves as needed, but it also means that they have stocks to deplete. In some cases, there is real value in being able to tap stocks, which may be depleted, but are also renewed at a rate that is somewhat comparable to the rate of use. Dam-based hydropower is such an in-between, or combination, resource. Stored potential energy can be depleted, but they tap substantial natural energy fluxes that can be used sustainably. Likewise, tidal power has not yet been significantly tapped, but it may have very similar resource characteristics to prevailing large-scale hydropower systems. It may impound water at high tides, to be available throughout the day, creating its own stock.

Geothermal energy's reserves are accumulated by geologic processes as tectonic activity in the Earth's crust. Current geothermal projects rely on accumulation of heat in reservoirs, of porous or naturally fractured rock. The pores are water-filled and that water is superheated by the heat from the Earth. When that steam is produced through a well, it can be used to drive turbines to generate electricity. As steam carries heat to the surface, the heat reserves are being depleted. As the temperature of the reservoir is drawn down, heat flows in from underlying and surrounding rock. This constitutes a renewal process, which extends the value of the resource. Nevertheless, if our depletion exceeds the renewal rate, it will become limited.

Most biomass resources that humanity currently uses tap into stocks that are intrinsically limited and can be exceeded. Although viewed as renewable, they are constrained by their limits. Forests are directly depleted by the levels of use in many fuelwood-dependent regions. The depletion of forest reserves is a double-faceted cost. Not only are stocks removed by overconsumption, the loss of forest also reduces the renewal rate. (There is an additional cost associated with the depletion of forest resources related to the loss of ecosystems.) Other biomass resources have similar negative values, although generally less direct. Corn ethanol requires land and water dedicated to growing the biomass crops. It is not an entirely natural flow of energy that is tapped for biomass crop fuels. Human activity tills the soil, fertilizes, irrigates, and plants crops. In order to do these things, humans clear the land, removing the vegetation that was growing naturally. The source of the source (the land and water and nutrients streams to grow the crops) can be depleted. There are definitely doubts to raise as the sustainability of biomass cropping. A likely exception may be growing algae, which does not take up nor deplete crop land, and can use waste streams for its nutrient sources.

The resources that draw on purely natural energy flows are generally drawn from solar energy. Obviously solar, but wind also, which is a secondary product of the uneven heating of the Earth's surface by the sun. They both carry vast amounts of energy to and on Earth. The intensity (energy density) is not very high though, especially when accounting for the variability of energy flows. The availability of solar energy drops to zero at night. It declines markedly during winter months and when the skies are cloudy. To have energy available on demand with these systems requires storage. Those systems that have no natural stockpiles need engineered storage.

Finally, nuclear power is another energy alternative that has stockpiles that are tapped for energy production. Indeed, nuclear fuels are probably the most absolutely nonrenewable, as there is no Earthly regeneration process. The great value is the immensity of energy that can be released by fission processes. Primarily uranium is accumulated in veins or beds that can be extracted by mining operations, or in situ leaching into wellbores. The mining or drilling and leaching processes hold similar risks to other mining and drilling operations. However, a pound of purified uranium 235 produces the same energy as more than 5000 barrels of oil per day or 1250 tons of coal per day. So, the energy resource base is very high, even though the raw material is quite rare.

Final Thoughts

Resource value is conspicuously tied to the abundance of that resource. The abundance of stock-limited resources decreases with time, as we produce the stocks. This is true for all of the fossil and nuclear fuels. It is impossible to know precisely what the total stock (or resource base) is, but do not be deceived, it is finite and is depleted by production over time.

Availability of resources is equally critical. The heat of the Earth's core is immense, but the core is totally unavailable to existing technologies. Extensive exploration is required to make available fossil and nuclear resources that have not yet been discovered. Part of the limitation of flow-limited resources is that their availability is limited not only by the flow rate of the source, but also by the location of the energy flows. The energy produced by these systems cannot be readily stored or transported over long distances. The value of every energy system depends on the ability to make energy available to the people who need it.

Note that both the availability and abundance factors change over time, as resources are depleted and new resources are made available. There are also time factors in terms of how long it may take for new energy systems to develop technical viability. An abundance of energy can be drawn from tritium and deuterium, but none of it adds to our overall energy system values until the technology for controlled fusion is developed and disseminated.

References

Daly, H. E. 1994. For the Common Good: Redirecting the Economy toward Community, the Environment, and a Sustainable Future (No. 73). Boston: Beacon Press.

Deffeyes, K. S. 2010. When Oil Peaked. New York: Hill and Wang.

Energy Information Administration (EIA). 2014. "World energy outlook: Global petroleum and other liquids consumption." Accessed from: http://www.eia.gov/forecasts/steo/report/global_oil.cfm

Environmental Protection Agency (EPA). 2014. "Carbon pollution standards." Accessed from: http://www2.epa.gov/carbon-pollution-standards

Grossling, B. F. 1977. "A critical survey of world petroleum opportunities," Project Interdependence, Library of Congress. Committee Print, 95–33, 645–658.

Horowitz, D. 2005. Jimmy Carter and the Energy Crisis of the 1970s. New York: St. Martin's Press.

Humphreys, M., Sachs, J., & Stiglitz, J. E. (Eds.). 2007. Escaping the Resource Curse. New York: Columbia University Press.

Maggio, G., and Cacciola, G. 2012. "When will oil, natural gas, and coal peak?" Fuel, 98, 111–123.

Masters, G. M. 2013. Renewable and Efficient Electric Power Systems. Hoboken: John Wiley & Sons.

Schölnberger, W. E. 2006. "From scarcity to plenty. Who shapes the future mix of primary energy? What might it be?" Oil and Gas European Magazine, 32, 8–20.

Smil, V. 1987. Energy, Food, Environment: Realities, Myths, Options. London: Oxford University Press.

Smil, V. 2006. "21st century energy: Some sobering thoughts." OECD Observer, 258, 22.

The New Times. 2013. "Uganda wanes off foreign pressures as oil production nears." Accessed from: http://www.newtimes.co.rw/section/article/2013-04-20/65066/

Watkins, G. C. (2006). "Oil scarcity: What have the past three decades revealed?" Energy Policy, 34(5), 508–514.

World Alliance for Decentralized Energy (WADE). 2014. "Mission statement." Accessed from: http://www.localpower.org/abt_mission.html

Yergin, D. 2011. The Prize: The Epic Quest for Oil, Money & Power. New York: Simon and Schuster.

CHAPTER 3

Value Added and Lost in Energy Services

Energy as a Service

Although energy is essential to all human activities, it does not have value in its own right, but only in use to support services and products. Energy brings value as we convert it to end use, and some uses offer more important benefits than others. On one hand, cooking our meals has been an important step in human development, becoming a practical survival need. On the other hand, running an air conditioner in a car parked in a shopping center parking lot actually offers exceptionally little benefit, for a large amount of energy consumed.

Indeed, a great deal of energy we use does not serve any useful purpose. Inefficiency tremendously decreases the value of energy use. The end use efficiency of automotive transportation, for example, is less than 20 percent, but it is even less when considering the total system and the outcome sought: moving individual people. Thus, it is imperative to understand what we want from our energy systems, in order to optimize the benefits we get from energy use: what forms of energy serve our needs most effectively. Every kind of energy demand calls for specific forms of energy.

Converting Energy to Societal Value

The rise of humanity has been a story of increasing energy use. It has also been one of increasingly sophisticated energy systems, with greater energy and power densities, more readily transported and dispatched. At its core, the energy–society linkage involves three essential points:

- Energy serves basic needs for human survival

- Energy is an essential building block to all development activities
- Needs are specific in terms of when and how energy is required.

Each point represents a substantial investment in services and infrastructures that must be dealt with as a whole when valuing how each resource fits or can fit going forward.

Matching Energy with Needs

To gain the most value from our energy consumption, we must use energy that is in a form appropriate to the purposes it serves. Electricity is the only form that can power electronics: phones, computers, televisions, and technical instruments. While electricity can be used for cooking (those of us with microwaves certainly appreciate that), it does not readily replace firewood or charcoal for cooking. It requires different utensils, and it does not produce the same flavor, or cook the same way. Do not be fooled into dismissing flavor as a purely subjective, discretionary value. Taste is, for all creatures, a means of identifying that which is or is not good to eat. New tastes can be learned, but that does not come automatically.

Consider the story of a solar cooker project attempted in Western Africa, as told to us by a UN official. Well-intentioned scientists from the West knew that it is possible to use parabolic reflectors to concentrate sunlight to generate intense heat. They delivered a large number of free parabolic solar cookers to rural households in Western Africa. The scientists brought along packages of hotdogs to demonstrate how the cooker worked. Upon returning a few years later, they found that the only cookers in use had been filled with charcoal, a grate placed over the top, and were used for barbecuing. Why? It turns out that the rural villagers were agrarian: all of the adults worked in the fields from dawn to dark and did all their cooking at night. Further exacerbating the problem, it turns out that hotdogs were not the staple food in this community and the more common stews did not fit well on the skewer in the focal line of the parabolic reflector.

Even in the United States, where most of us do not work in the fields until sunset, solar cookers are not popular. Why? Partly because most people place a high value on the convenience of being able to cook what they want when they want. The *value added* by an energy system is in part a function of how well it fits the usage patterns of the consumer: an issue of social acceptability.

Necessary and Sufficient Energy

For the world's poorest, a little more energy correlates to a vast improvement in quality of life (Martínez and Ebenhack 2008). While such a correlation does not prove causation, the correlation does make logical sense. Energy does serve essential needs and it supports every step of development. Imagine being stranded on a deserted island. What would you want to be sure you had access to? Food, fresh water, shelter, and energy, in the form of fire. People will walk miles each way to gather the firewood for their weekly meals, or they are willing to spend as much as one-third of their household incomes for it (WCED 1987). Without basic energy, basic survival needs cannot be met.

It follows then, that without more energy, basic development needs cannot be met. But even energy, alone, does not cause development. There must be a confluence of development opportunities with the requirements to produce them. We suggest that energy is not a sufficient, but is a necessary ingredient to development.

As long as more energy continues to serve development needs that contribute measurably to quality of life, a positive relationship will persist, but beyond a point of sufficiency, the benefit increases more slowly. This is to say that more energy continues to help to improve quality of life, but not as dramatically. We suggest that this is the point at which the most critical survival needs are met. People have sufficient energy not only to cook their meals, but to provide for food preservation and delivery, not to mention access to pure water, where life expectancies begin to increase dramatically.

Increasing access to energy occurs not only in amount, but in kind as well. Raw biomass is a limited and limiting resource. The transition beyond minimal energy needs in modern population levels is accompanied by increased access to more modern energy resources. This is essential in order to support all of the other development activities. Hospitals, schools, manufacturing, and transportation do not run on firewood—let alone on dung. Fluid fuels and electricity are absolutely necessary to support these activities. The step up the "energy ladder" to more modern energy systems also means safer, cleaner uses.

Progress continues beyond the point of sufficiency. Let us liken energy to food. Not only are both essential to surviving and prospering, but also have sufficiency levels and saturation levels of consumption. Being below the sufficiency level of food consumption is referred to as malnutrition. It leads to many health problems. Having enough food to be adequately nourished does not mean that you are getting all of the benefits of an abundant diet; it just means that you have enough to survive. More (and higher quality) food can promote robust physical activity and mental development—and is certainly more satisfying than having barely enough. Beyond the sufficiency point, though, there are no more real benefits of consuming more food. Indeed, beyond some point, quality of life actually declines with greater food consumption. We get fat, have high cholesterol, run increased risk of diabetes, and have various other physical problems.

We suggest that, like food, once energy passes the threshold of sufficiency, more energy continues to support further development, but much more gradually. It supports industry, commerce, and public services. Schools need lights. Hospitals need lights, refrigeration for medicines, and power for sterilization and instrumentation. All of these things help us to live longer or better. Once all of these needs are met, more energy no longer adds real benefits, but using more energy does incur all of the risks and costs of energy use, without significant enhanced benefit; so we say that a saturation point has been reached. Curtailing excessive energy use back to the sufficiency point may be beneficial.

Energy Conversions in Society

Our use of energy almost always involves some conversion of energy from one form to a more useful form. And, generally, there is a conversion at the point of use. Among the most basic conversions, is burning fuel to

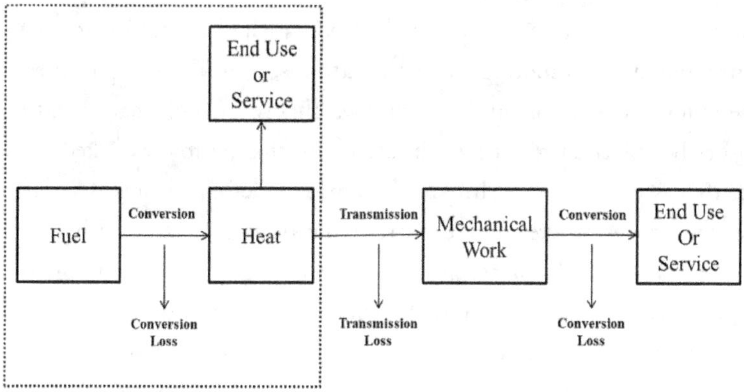

***Figure 3.1** Box model representation of energy conversion processes. The dotted area represents direct use after conversion, such as in cooking or heating.*

produce heat. That is, converting the energy stored in chemical bonds to heat. This conversion is commonly used for cooking and for space heating.

Running machinery calls for a second step, in which the heat causes gases to expand, to push a device such as a piston or a turbine to produce kinetic energy in the machine. Very commonly in modern society, that machine is a turbine, whose rotation generates electricity. In this case, we have converted chemical energy to kinetic energy and in turn to electrical energy. The electricity serves as a particularly versatile "energy carrier" (see Figure 3.1).

Inevitably energy is consumed in the midstream processing (heat loss, the primary culprit), resulting in reduced overall system efficiency. Naturally, there is pollution and other environmental impacts associated with each step as well. Money is spent as well, although those expenditures can contribute to jobs and other economic benefits. Other values of midstream conversions are related to the improved dispatchability, energy density, and cleanliness of the resultant energy form.

Process Heat

As described above, all kinds of processes require heat. The simplest and most ubiquitous is cooking. In this process, food is transformed by heat:

changing its flavor and texture; killing potentially harmful microorganisms; and even promoting chemical reactions, as in the rising of bread. Certainly, there are many foods that we choose to cook that would be just as beneficial raw, but most meats are much safer to eat cooked. Water that is boiled before drinking eliminates microorganisms that carry deadly diseases, where community scale purification is not available.

Survival outside of tropical climates is largely dependent on the availability of space heat. Comfort requires even more space heat. Combustion fuels continue to provide for a vast majority of space heating needs. Additionally, most manufacturing involves some processes that must add heat. All metal working calls for heat to smelt the ore to purify the metal. Then, it must be shaped and formed.

Illumination

Taken for granted in affluent societies because of its ubiquity, lighting the dark has been one of the most important values that humans get from energy. We can now have light any time with the flip of a switch. But, imagine the significance of artificial light to early humans. Fire light provided safety and allowed people to work after dark. We suggest that is really how innovation flourished as humanity began to rise above our competing species. People could hunt and gather food through the day and sit around the fire at night, experimenting with ideas for primitive tool making.

In the past couple of centuries, the ability to have light allowed children in farming families to read at night—transforming literacy rates throughout the Industrialized World and allowing a new kind of social mobility. One no longer needed to be born into affluence to have the luxury of education. This allowed for rapid growth in innovation, as far more people were able to shift from physical to mental labor. The limitations on energy for illumination still constrain literacy and education in many Developing Countries. Even at residential schools, students in poor regions are often limited in their ability to study in the evening by the lack of lighting.

Fire was the basis for lighting, as well as cooking and heating, until quite recently. The technology of combustion to produce light improved with fuels that burned relatively clean and bright over the millennia. Coals in braziers or torches permitted bringing the light into buildings. Candles and oil lamps, using oil derived from vegetables or animal fats, were much brighter and cleaner. Candles and oil lamps permitted reading, which had been very problematic with only wood, charcoal, or torch fires.

The advent of kerosene made another large step forward. It was a safe, convenient, bright light. Even today, many people who enjoy the convenience of electric light keep some kerosene lamps as backup in case of a power failure. Within a couple of decades of the new petroleum industry bringing kerosene to the market, the co-produced natural gas was found to be superior to town gas and available in large quantities, with relatively little processing.

Electricity brought the largest leap in value added through illumination. It was not explosive or toxic and had no combustion products released into living spaces. It was capable of creating bright light, with a wide variety of qualities and colors. In modern, affluent societies, light surrounds us and enables us to work and play comparably well in day or night.

Incandescent lighting, a result of glowing filaments that resist the flow of electric current, has been the dominant lighting technology for over a century. The inefficiency for which the incandescent bulb has become notorious stems from the fact that it generates more heat than light. Fluorescent lighting is much more efficient, in which electrons pass through an inert, ionizing gas. Fluorescent bulbs commonly contain trace amounts of mercury, which fluoresce when bombarded by electrons, and then activates a phosphor coating, which then emits light in the visible range of the light spectrum. Though a much more complex process than incandescence, the end goal is less heat generated for equivalent lighting.

Even more efficient than fluorescent lighting is the light-emitting diode, or LED. A diode is a junction of positively to negatively charged materials. In the case of LEDs, there are "holes," which passing electrons may fill. As they do so, they emit energy. Different materials emit light in different wavelengths, so selecting the materials carefully permits design of devices producing virtually any color of light desired, including white light. The diodes themselves are extremely small, so LEDs lend themselves to use in displays, as well as in large arrays to produce area lighting. LEDs are not only more energy efficient, but they have much longer service lives than either fluorescents or incandescents. They do not have any flicker and can be dimmed easily. They turn on instantly. They truly have all of the advantages of incandescents and of fluorescents, except for low cost. It seems likely that LEDs will take over most of the lighting market share as economies of scale bring down their manufacturing costs.

Mechanical Work

The most obvious form of work to get from energy is mechanical work. Of course, before humans were able to tap into external energy sources, we were limited to the physical or mechanical work that we could do ourselves. Around the time of established agriculture, humans began to domesticate "beasts of burden" to supplement human labor. Horses, oxen, water buffalo, camels, and elephants are much stronger than humans and their work was able to expand our productivity greatly. Animal power, whether our own human labor or that provided by larger domesticated animals, was the primary means of doing work for a million years of human prehistory.

Grinding grain was one of the first tasks to use external, nonanimate energy sources to augment human or animal labor. This laborious, repetitive task was well suited to the applications of water and wind power. The utilization of gearing allowed the rotational motion imparted to a wheel by flowing water to be translated into rotating a large grist wheel in a horizontal plane rather than the vertical plane of the waterwheel. (Applying similar principles to wind simply required the innovation of putting sails on arms extending from a wheel.)

The same sorts of rotational movements, with the application of a crank system, could be turned into reciprocating motion. This innovation allowed for a wide range of applications in lumber mills and then textiles. Much of early civilization grew up along rivers not only because of the transportation benefits, but because mills produced so many useful goods.

Many pumping systems also utilized the conversion of rotational to reciprocating motion. Pumping water remains essential to having clean water. A recent survey in the *British Medical Journal* showed that of

11,300 respondents, sanitation was found to be the greatest medical advancement since the journal began publication in 1840 (Ferriman 2007). Hence, this certainly speaks to the great value derived from being able to move and treat water effectively. Pumping water, though, was also pivotal to the ability to carry out extensive underground mining, to extend irrigation for agriculture, to settle dry lands, and even to "reclaim" land from the sea, as is done in Holland.

The sophistication of manufacturing has continued to advance, making a wider variety of more minutely engineered products, still largely based on rotational and reciprocating movements.

Transportation

A contemporary form of widely utilized mechanical conversion is transportation. We discuss it separately because it is such a dominant energy consumption sector in the world economy, but in the Developed World in particular. Consider, for example, that approximately one-eighth of the word's oil is consumed by the United States transportation sector. In other words, 5 percent of the world's population consumes about 12.5 percent of the world's annual oil production in this sector alone.

Much of transportation consists of causing gasoline or diesel to explode within a cylinder, where it drives a piston the length of that cylinder, which turns a crank, which provides rotational motion, which then turns the driveshaft. This tried and true technology remains the main driver for all of propulsion today. Any combustible fluid can, at least hypothetically, be used to drive an engine.

Before liquid and gaseous fuels were broadly available, solid fuels were often used to generate steam, which could also drive pistons to produce rotational motion that drove many factory processes, steam locomotives, and paddlewheel boats. The process worked, but at extremely low efficiency. Its application to transportation was limited by the dispatchability of fuel, since the solid fuels are bulky. Some was stored onboard, but entire rail cars are needed to store wood or coal for locomotives.

The energy densities provided by fluid fuels enable them to carry a great deal of energy in relatively small volumes and low weight. This improved the efficacy of waterborne carriers and opened the door to flight. The first successful airplanes also used internal combustion engines to turn propeller blades. The advent of jet engines allowed for much higher speeds.

Placing Value on Conversions

Energy conversions are essential to gain the values we seek, but the conversions come at their own costs. Naturally, every conversion process requires economic investment. Just as naturally, each step has efficiency costs. Every transformation loses access to some of the original energy value. Furthermore, as with every human activity, every step of the conversion processes produces some waste products and some social impacts.

The aggregate value from the conversion processes is the sum of values that the energy conversion offers at its end use, minus the inefficiencies and other costs associated with each step. The fact that considerable value is added by energy in making electricity available to consumers, in manufacturing goods, in supporting critical development activities, in lighting, warming, or cooling our spaces and transporting us and our goods is undeniable. It is arguable whether—and how—energy by-products should be accounted for.

Additionally, there are some basic tenets regarding the viability of an energy source. Energy's value is in its usefulness in performing work. If it does not work, it is not an energy source. Also, any energy system must be evaluated based on the net useful energy it provides to humans. Finally, thermodynamics limits the viability of energy technologies and can be summarized: you can't win; you can't even break even.

Indeed, the value of any energy system is predicated on its energy production. We argue that value should always be evaluated as a function of the useful energy, or work produced. Of course, the fossil fuels generate not only the greatest environmental and social impacts, but also, by far, the most useful energy. And useful is the operative word. Energy is only of value insofar as it serves human needs or interests. Only the energy that adds to what we may use is beneficial. If any nonfossil fuel energy system were to have externalities in any way comparable to the fossil fuels, it would be of no value, since those costs would be associated with so little useful energy.

Infrastructure

One of the costs of establishing an energy system is the infrastructural support required to produce the energy, to deliver it to end users, and to consume it. Even the handling of waste products requires infrastructure. Oil and gas require pipeline systems and refineries are as necessary as the wells. Coal requires rail and barge transport to move it from the mines to the manufacturing and power plants. When any source is used to generate electricity, extensive power lines are required to get it to businesses, institutions, and households.

The term infrastructure refers to everything that is indirectly required to support an activity, but infrastructure can be considered to go much further. Every energy industry depends on significant ancillary industries. Oil and gas production depend on service companies, which may be larger than many of the oil companies they serve. These ancillary companies provide drilling fluid (i.e., mud), log the wells, provide steel casing and cement, perform the fracturing, and even drill the wells. Solar and wind power depend on mining and processing industries to make the semiconductors and the electronics. Large-scale hydropower requires monstrous quantities of concrete and steel.

Consumption activities have their own infrastructural requirements. Transportation is one of the most significant energy consumption sectors and one that entails tremendous infrastructural demands. It requires vehicles and roads on which to drive those vehicles. Beyond that, the vehicles need fuel. In order to support travel all across the country, there must be many fueling stations, spread out to accommodate the needs for more fuel, wherever they occur along journeys. Thus, extraordinary infrastructure is needed: with thousands of service stations and with pipelines, storage terminals, and many thousands of trucks to deliver millions of gallons of fuel to those thousands of service stations every day.

The extensive infrastructure established in affluent nations in support of petroleum-based fuels is essential to making them cheap and abundant. If fuel were not abundantly available, it is quite clear that transportation would not be viable. It is less conspicuous, but no less true, that transportation would not have developed as it has if fuel were expensive.

Combustion fuels have extra value in their ability to be stored and transported. Energy-dense portable fuels were essential to the growth of transportation, which expanded rapidly as fuels with greater energy density became available. Coal replaced firewood and was, in turn, replaced by petroleum products. In the early days of locomotive transportation, nature provided a form of infrastructure in the shape of forests, from which firewood could be harvested.

Railroad locomotives carried some of their own infrastructure: coal or firewood cars to store enough fuel to have some range. With the coming of the internal combustion engine, much greater mobility was permitted. Automobiles carry enough fuel to cruise a few hundred miles, but need to be able to refuel along the virtually unlimited routes, necessitating many thousands of fueling stations.

If society transitions toward other kinds of energy for transportation, the new sources will need to be comparably widely distributed in fashion similar to the gasoline and diesel fueling stations. The new transport fuel infrastructure must also be capable of meeting refueling needs in a timely fashion. Electric vehicles store energy in massive batteries, which could take hours to recharge. Technologies for quick charging are still not likely to be as fast as motorists have come to expect from energy dense combustion fuels. It may be feasible to swap out depleted batteries, leaving plenty of time to recharge at the service station. Such a scheme can even readily employ solar or wind power to recharge—mutually supporting noncombustion energy alternatives and alternative, low-emission vehicular power systems. Since the energy for vehicular use must be stored anyway, the process of charging batteries is not an added expense of efficiency loss; thus, the intermittent nature of solar and wind become insignificant in this application.

Public Utility

The entities providing essential services to people are commonly referred to as public utilities, which include electricity, natural gas, water, and sometimes telecommunications. In particular, the public utility requires a great deal of infrastructure to generate energy, transmit it to communities, and deliver it to homes. As these services for electricity and natural gas first began to evolve around 1900, competing companies built their own infrastructure and tried to sell competitively (US DOE 2002).

There were inherent inefficiencies in building multiple, competing infrastructures. Monopolies emerged, which overcame the inefficiencies, but prompted public concerns for the potential of price gouging. Regulated public utilities followed. In many cases, the utilities were regulated by an oversight board or commission, with guidelines that established fixed profit margins. Prices were established to match the selected profit margins. The costs of operating are figured in, then the selected profit margin is added and that sets the price to be charged to the customers.[1]

In the United States in the 1980s, President Ronald Reagan, a strong believer at that point in his life in the advantages of free market competition, pushed for national deregulation of the public utilities. There were certainly serious problems with the early transition to "free market" forces. Although a number of questions had been raised about improved value to the consumer in Reagan's time, arguments now are being advance that the debate seems to favor deregulation (Pentland 2013; Slocum 2001). Either way, there remains room for debate on how certain essential services are delivered to the household and whether or not that may be better delivered by regulated public utilities. (This will certainly be exacerbated as more people begin adopting grid-connected, local generation on their rooftops.)

Whether tightly regulated or not, the utilities are expected to deliver sufficient energy, on demand, for the consumer. This requires sufficient investments in production, storage, transmission, and delivery infrastructure. In the Affluent Nations, the availability of natural gas and electricity, on demand, is commonly taken for granted, but the same cannot be said in the Developing World. Rolling brownouts, somewhat scheduled shutdowns of parts of the grid, are common. Couple this to

---

[1] It seems that the public perception of the petroleum industry is that it should operate as a public utility: that the price of gasoline should be fixed, based on the costs of operating. Whenever gasoline prices go up (in response to market forces), there are commonly questions raised by consumers and reported by the media about why, when the cost of operating did not go up that much? That's free market economics, as opposed to a public utility – prices are set by market forces, not by operating costs.

not always having sovereign control of its own power grid (due to controlling rights by outside investments) and it all but guarantees an insufficiency of the local grid to maintain power. It interferes with, rather than supports, many development activities.

Future Infrastructures

What is the most effective form of infrastructure to support development going forward? The current model of national grids, extending across continents, seems in many discussions to be the model to pursue. It is built to serve a very large scale of production and consumption. However, national grids require immense investment to establish and large sums to maintain. For people consuming only a tiny fraction of the electricity that westerners do, economies of large-scale production cannot be as readily accomplished, without needing to transmit the energy very long distances. Long distance electricity transmission inevitably incurs transmission losses, meaning that more fuel must be consumed, to produce more electricity, more of which is lost in transmission.

Numerous development projects have been proposed for grid extension, in order to reach more of the population. We analogize this to a dendritic model of electric grid growth. That is to say, it starts at a central point and spreads out in all directions, like the dendritic roots of a plant spread out from the stem. We recommend, instead, what we like to call a nucleation site model. This term derives from the manner in which raindrops form around numerous nucleation sites in clouds (generally dust particles).

We argue that this is more reflective of the manner in which electric grids formed in the Developed World. Only after many nucleation sites had developed across nations did they begin to reach out to one another to interconnect. Whereas the overall cost per gigawatt hour produced may be lower for large-scale power plants, the individual capital investment required for each power plant is much lower for small facilities. This can be a big difference for Developing Countries, with very limited access to capital.

The nucleation site model is also more amenable to local production of flow-limited resources. Solar photovoltaic energy production is extremely well suited to local, distributed production. It is also extremely modular, which makes it scalable, from a very small scale, upwards. Panels can easily be mounted on the rooftops of the homes and businesses using them. Generally, wind power has been developed at larger scale—still many times smaller than the gigawatt scale of commercial coal and nuclear and hydropower plants—but can potentially be placed on rooftop, or even urban lots.

Multiple consumers can be linked together to form a microgrid, by which individual generation sites can support one another, leveling production and meeting demand variability between the two consumers. Once energy is being generated to meet local needs and expands throughout a community and the same thing happens in a neighboring community, a point can be reached at which two communities can be linked to form a larger nucleus. We suggest that this sort of strategy offers greater value, in terms of likelihood of success and of sustainability. Modest investments in a number of communities can provide energy to those communities, without the major nationwide infrastructural investments and it will promote the use of more sustainable energy sources.

Hydrogen Infrastructure

At the beginning of the current century, there was a great deal of enthusiasm for the emerging "hydrogen economy." The promise was that hydrogen in fuel cells could power almost anything, especially vehicles, with no harmful exhaust emissions at point of use (see Rifkin 2003). The hydrogen itself must be viewed as an energy carrier, as there are no sources of elemental hydrogen on Earth and more human controlled energy is required to synthesize it than it contains.

The value of hydrogen systems would be primarily in terms of very clean energy production at the point of use and the possibility that the hydrogen could be generated by a clean alternative, such as solar or wind energy. Indeed, hydrogen fuel cells could provide the storage to offset the variability of solar and wind power. Under this scheme, hydrogen would be generated during off-peak hours by electrolyzing water to yield hydrogen and oxygen. Then, in a fuel cell, the hydrogen and oxygen could be recombined, yielding water and electricity.

If the hydrogen economy is to become a successful part of global energy systems, it will require new or significantly modified infrastructure. Steel pipelines and storage facilities that hold oil and gas are not suitable for hydrogen. Problems can be experienced either as embrittlement (in which hydrogen enters submicroscopic pores or voids in steel and recombine into diatomic hydrogen molecules, creating pressure that can initiate cracks) or as hydrogen attack (in which hydrogen reacts with carbon in high carbon steels to form methane, thereby removing carbon from the alloy.) There are materials, including steel alloys, that can hold hydrogen, but it will commonly not be the same as what currently exists.

If hydrogen fuel cells are used in vehicular transport, with high pressure storage, there will have to be storage and transfer facilities to handle those high pressures, as well as the onboard storage vessels. Wherever the hydrogen is generated, it must be transported to the refueling stations. This likely means high pressure compression equipment at every fueling station, or transport at extremely high pressure. The latter would mean that the entire distance of the hydrogen transmission network would operate at pressures that would be extraordinarily dangerous if any leaks occurred.

Either way, the fuel would have to be dispensed at the extreme pressures, which would make "self-serve" customer fueling challenging. Any mistakes could present a real hazard if it escapes within an enclosure (like the roofs that are popular over many fueling stations), because the hydrogen is extremely explosive if it accumulates. Therefore, we suggest that a hydrogen economy for transport will necessitate a return to full-service fueling stations, at least in the beginning. The costs of all of these infrastructural components would surely add to the overall cost of the system.

Electric Car Infrastructure

A related option to fuel cell cars, and one that has more likely near term potential, is the electric vehicle. Pure electric cars will require infrastructural changes as well. Batteries will need to be recharged frequently and in a timely fashion. Motorists have come to expect that they can drive practically unlimited distances, stopping only for a few minutes to refuel every few hundred miles. Technology that permits very rapid charging of batteries could address this problem, but there are other issues. For example, such rapid recharge rates may generate a great deal of excess heat. It also means that much more grid power will be required at any given time.[2]

Whichever technologic approach that gets adopted will require radically transformed service station infrastructure. Either system will require standardization. Batteries must be interchangeable or the charging characteristics must be uniform. Fueling stations are already ubiquitous in much of the Developed World, but the maximum distance between charging stations is likely to need to be limited.

Roads, Rail, and Air

In the years following World War II, President Eisenhower's administration undertook a fateful decision to set the United States on a path to its transportation infrastructure based on extensive surface roads. This decision also had the effect of promoting the private automobile. This, in turn, promoted the radical growth of suburbia.

The transition to surface roads and private vehicles presented a nearly insurmountable challenge to the railroad system in the United States. As energy constraints raise the stakes on conservation, expanded public transport is likely to be part of the solution. That will push the United States to redevelop rail systems that we spent much of the past 60 years abandoning. If done, the new rail systems for human transport will predominately be for high speed rail. This will allow rail to compete with short-to-medium haul air transport.

Although it does not need roads or rails, air transport has its own infrastructural needs. Airport complexes are massive structures, but merely the tip of the iceberg. There are thousands of radar tracking stations involved to monitor the many thousands of daily flights, across the globe and for short regional hops. Fueling new air infrastructures also might well involve fuel cells and high-capacity batteries, but new materials for creating lighter aircraft would be more advantageous in the short term.

---

[2] A Low-tech Magazine (2009) article estimates that an electric car with a 24-kWh capacity battery pack requires grid power output of 3000 W for an 8-hour charge, but 144,000 W of grid power output to charge the same car in 10 minutes. For one battery!

The Refinery's Role in Future Infrastructures

Petroleum processing in refineries is vital to transform crude oil from a very low-value fuel into very high-value fuel. The potential value has, as we have said, been enhanced greatly by the millions of years of accumulation, concentration, and intensification through geologic processes, but most crude oils still have a wide variety of organic constituents. The petroleum industry took off based on a relatively simple separation of moderate range molecules to form a product that was called kerosene.

Kerosene burned bright and clean, making it an excellent fuel for lighting. As gasoline-powered vehicles came into prominence, gasoline (comprising a lighter set of components) became the desired product. Kerosene, though, remained valuable for niche heating and lighting applications and diesel utilizes a similar, but broader, range of constituents.

Experimentation found more and more products of value that could be derived from the broad organic chemical content of petroleum. An incredible array of synthetic products are currently made from petroleum, accounting for approximately one-third of the volume of the average barrel of oil, alongside the array of versatile, energy-dense fuels that are used for every form of transportation. We are all surrounded by petroleum products in the Developed World. Thus, petroleum and its products are likely to have value as petrochemical feedstocks even when noncombustion alternative fuels have taken over the marketplace. We will still likely use rubber tires, petroleum lubricants, and asphalt roadways in any future scenario.

Final Thoughts

Energy provides essential support for human survival and development. Generally, though, energy serves us indirectly, through conversions to do work. Manufacturing, transportation, communications, cooking, pumping and purifying water, and preserving foods are all services that are based on energy.[3] We are surrounded by energy. Every human activity is based on using energy. Most of these services depend on converting energy from one form to another.

---

[3] Space heating and lighting are probably the exceptions in which the energy itself is the service.

Delivering the energy to the point of use at the desired time of use requires storage and transmission systems: energy carriers. All of this, in turn, requires supporting infrastructure. One of the challenges faced by the emerging alternative energy systems is that they are competing against energy systems with fully developed infrastructural support systems.

Valuing energy appropriately must be based not merely on the size of a resource base, or the environmental costs, but on the value of the services provided. As energy is converted to more versatile or useful forms, there are costs and efficiency losses with each step. The network performed by an energy system is the basis of the value added by the system.

Petroleum valuation would benefit the most from inclusion of byproducts, as it serves as such a rich organic chemical feedstock. However, there can be no question that large-scale fossil fuel power plants and refineries are not particularly attractive. They are also extremely large-scale facilities, which means that they incur some worker hazards.

Although hydropower facilities tend to be rather clean and hidden within the massive dam structures, the dam that impounds water to provide a constant, intense flow of water through the turbines has its impacts. They do disrupt ecosystems and they often displace residents.

Wind power is joining hydropower as a non-polluting, non-depleting energy system that is operating at increasingly large scale. Indeed, as wind farms have begun to operate at utility levels, with relatively large power plant footprints, they are beginning to carry a set of environmental and social costs that seems to have surprised many people. (Perhaps energy systems are adorable in their "puppy" stage, but have challenges as they mature.)

Large-scale solar power facilities are not likely to be very hazardous, nor likely to require many workers to maintain. The manufacture of many billions of photovoltaic cells will be an extremely large-scale activity, which will include extraction of minerals and handling vast amounts of materials, some of which are toxic. We definitely expect solar power to be cleaner and safer than combustion fuels, but want to caution that such large-scale activities must come with their own hazards.

Nuclear power is no doubt safer and cleaner than is commonly perceived. Radiation is no more dangerous to humans than chemical toxicity, explosion hazards, or serious physical injury. Still the hazards are real. Accidents do happen and some people are impacted by nuclear hazards.

A yet externalized cost of nuclear power in the United States is the deferred decisions on how to dispose of high-level radioactive wastes and of how to decommission nuclear power plants. Interesting proposals for single-use nuclear power plants that would be buried underground may mitigate some of the concern and actual hazards, especially for implementation in Developing Countries. Their effectiveness remains to be seen.

That said, we must temper excessive optimism. Besides the impossible dreams (think of the free energy from water or the super-efficient carburetor schemes), many real-energy systems are viewed naively, attributing more potential than is technically justified. Once again, the failure tends to be one of ignoring what all is required for an energy system. One of the prominent dreams today is that fossil fuels are obsolete and will be readily displaced by some "green" energy, probably solar. It ignores the technical limitations of tapping diffuse, flux-based energy sources to meet energy demands, which are often intensive and out of synch with solar flux. The dreams also ignore the tremendous inputs of materials and of energy to build and deploy "big solar."

We do believe that solar energy will continue to develop and is likely to become a serious contributor to the evolving global energy mix, but it is important to note two things. First, solar is not a serious piece of the global energy pie at the writing of this book; it provides well under 1 percent of the total. Second, because it is currently producing so little energy, it will require tremendous inputs: investment, materials, and energy to manufacture huge quantities of solar-based energy systems.

A variety of solar energy technologies are technically viable. That variety, though, represents its own challenge. Even as we reach some consensus that solar power must grow. It remains unclear which technologies will emerge as champions of this transition. In addition to the photovoltaic cells, which most people envision, there are concentrating solar or (solar thermal) energy systems. Indeed, contrary to much common thinking, some of the largest solar power installations are solar thermal electric generation. A variety of concentrating solar technologies have been demonstrated to be successful: solar troughs, solar chimneys, and valleys lined with mirrors.

One of the seldom discussed advantages of concentrating solar power is that the variability of solar flux can be mitigated by the thermal mass of the working fluid that has been heated by the sun, which can support continued production after the sun sets. The most efficient photovoltaic systems, only operate in direct, intense sunlight. Some of the less efficient photovoltaic cells, though, such as amorphous photovoltaic arrays have an offsetting advantage of being able to produce electricity in diffuse light.

An advantage to all of the photovoltaic systems over concentrating systems is scalability. Almost everyone in modern, technologic societies use extraordinarily small applications, in our calculators and other small electronics. The typical panels that most of us envision are about 200 W. A small house or even school in the Developing World may be well served by 5–10 of these 200-W panels. Once panels are installed, there is no real problem in expanding the system, with more panels.

For instance, Peace School in Uganda worked with AHEAD Energy, a nonprofit energy development organization, to implement energy systems. With a 200-student school, and some residential properties, AHEAD installed 1.2 kW (that would be six 200-W panels). Once electricity production commenced on site, demand grew and it was easy enough to install another half dozen solar panels to double the onsite generation. Hence, photovoltaic systems can easily range from a small fraction of a watt, to millions of watts. The technology is viable, but we question the viability of the rate of scale-up needed to take a significant market share.

Likewise, wind power has developed much more rapidly around the start of the 21st century. It works. It can hardly be viewed as a novel energy form, since wind has produced energy for work for centuries. The modern wind turbines generating electricity build on the technologies of impellers (or sails), catching the kinetic energy of the moving wind and turning it into rotational energy: in earlier days to drive machinery; with modern wind turbines generating electricity. It is also scalable, both upward and downward.

Most residents of the Developed World have seen examples of the very large-scale wind turbines: reaching up to 8 MW of power for the world's largest turbines, with blades 80 m (about 250 feet) long (Wind Power Monthly 2014). Even with this towering scale, it would take 125 of these turbines to have as much capacity as a typical commercial size nuclear, or coal-fired, or hydro power plant. When you take into consideration the relative load factors, it would take about 1000 wind turbines to generate the same amount of energy. The challenge is not in the absolute viability, but in the technology and resources required to scale up sufficiently.

References

Ferriman A. 2007. "BMJ readers choose 'the sanitary revolution' as the greatest medical advance since 1840." The British Medical Journal. Accessed from: http://www.bmj.com/content/334/7585/111.2

Low-tech Magazine. 2009. "Who killed the electric grid? Fast-charging electric cars". Accessed from: http://www.lowtechmagazine.com/2009/03/fast-charging-electric-cars-off-peak-grid.html

Martínez, D. M., and Ebenhack, B. W. (2008). "Understanding the role of energy consumption in human development through the use of saturation phenomena". Energy Policy, 36(4), 1430–1435. Accessed from: http://dx.doi.org/10.1016/j.enpol.2007.12.016

Pentland, W. 2013. "After decades of doubt, deregulation delivers lower electricity prices" *Forbes*. Accessed from: http://www.forbes.com/sites/williampentland/2013/10/13/after-decades-of-doubt-deregulation-delivers-lower-electricity-prices/

Rifkin, J. 2003. The Hydrogen Economy: The Creation of the World-wide Energy Web and the Redistribution of Power on Earth. London: Penguin.

Slocum T. 2001. "Electric utility deregulation and the myths of the energy crisis." Accessed from: http://www.citizen.org/documents/MYTHS_Report.pdf, originally published in *Bulletin of Science, Technology and Society*.

United States Department of Energy (US DOE). 2002. "Electric utilities, deregulation and restructuring of U.S. electricity markets". Accessed from: http://www.pnl.gov/main/publications/external/technical_reports/PNNL-13906.pdf

World Commission on Environment and Development (WCED). 1987. Report of the World Commission on Development and the Environment: Our Common Future.

Wind Power Monthly. 2014. "The 10 biggest turbines in the world." Accessed from: http://www.windpowermonthly.com/10-biggest-turbines

CHAPTER 4

Economic Energy Value

The Economy: An Interconnected System

The global economy is an interconnected system that depends on resources to function. Thus, generally speaking, we can ascribe the following attributes regarding its role in providing social value via resource allocation:

- The economy allocates resources and provides incentives; it does not create resources.
- The economy does create incentives for innovation, which may solve resource problems, though not immediately.
- The marketplace cannot work for people who lack the means to purchase what they need.
- The marketplace does not capture all of the costs or benefits related to the production and consumption of resources.

Indeed, the economy is a means of allocating goods and services, and in a capitalist society, those with the greatest propensity to pay take the most resources. Propensity here is used to convey a combination of willingness and ability. Many simply refer to willingness to pay in resource allocation, which misses a crucial point, especially when taking a global perspective that includes the Developing World, where over a billion people live on the equivalent of a dollar a day. These people may have tremendous willingness to pay, which is rendered moot by their utter inability to pay for modern resources and services adequate to support a reasonable quality of life.

Some refer to the marketplace allocating oil resources, based on willingness to pay. For example, Tippee (1993) says, "in response to the consequently higher prices, consumers with the least intense needs for oil find ways to do without it." This, however, trivializes the reality that many people need more energy than they have the money to purchase. This explanation of the law of supply and demand typifies a faith in capitalism to allocate resources effectively, while fundamentally conflating the ability to pay with a willingness to pay.

Many people desperately need oil, but cannot afford the modern energy sources that they need. Does it make sense to say that there should be no concern over food shortages, because the marketplace will allocate food to those with the greatest willingness to pay? Or housing? Or clean water? All of these are essential to basic survival. It does not suffice to say that the consumers who cannot afford the essentials do not need them. Oil, or more generally, energy is an essential need for human survival.

The Marketplace

Inherent in the economy is the fact that monetary costs and benefits are the most apparent markers for valuing resources. No one wants to pay more for anything than is absolutely necessary. And everyone wants to get as much money or profit as possible. This tension is the heart and soul, as it were, of the marketplace and economics as a whole. From the perspective of energy resource allocation, the direct costs of energy procurement, delivery, and end use are highly structured and fairly easy to measure, only complicated by subsidies or rebates. Financial profits to companies are also relatively easy to track. The economic benefits to consumers, however, often are indirect, thus making it more difficult to evaluate.

Indeed, the marketplace has shown itself to work quite effectively at allocating resources and incentivizing investment and innovation in a well-functioning market. This means that well-informed consumers have access to goods that are available in surplus (or can be rather readily available). When some goods are not available, that represents a market failure. When the Titanic hit the iceberg, there was a tremendous willingness to pay for lifeboats, and, for many people, even a tremendous ability to pay. Sufficient lifeboats simply were not available.

While the metaphor may be trivial, the point is not. The market does not create goods or services. It creates conditions that offer incentives to produce those goods and services. That production still depends on the availability of the required materials and sufficient time to produce the products. And it depends on people having money to buy them. A properly functioning marketplace has several requirements. There must be sufficient goods or services to meet demand. There must be well-informed consumers. Of course, there must be an ability to pay. When these conditions are not met, the market has failed.

Just as the economy cannot provide what does not exist in a timely fashion, it assigns value based on the combined willingness and ability for people to pay. Since ability varies by several orders of magnitude,[1] those with the greatest ability to pay are spending a great deal of money on totally discretionary consumption (yachts and expensive wines and jewelry, etc.). This tends to drive up the prices on what is least essential, because that is where the richest are willing to spend their vast amounts of money. Thus, the monetary value of goods and services clearly do not reflect their essential values. If you were stranded on a desert island, would you rather have a diamond necklace, or food cooking over a fire? If the answer is not obvious, you have failed to imagine an urgent circumstance. The half of humanity that is constrained to severe poverty urgently needs even the most basic essentials, but does not have the money to reflect the value that these essentials have in their lives.

Furthermore, that for which a producer does not have to pay is generally not reflected in the market price of goods. Until quite recently, producers did not have to pay for polluting the air or the water. Nor did they have to pay for exploiting workers—even for fatalities in their mines or plants. In some of the less affluent lands, producers still do not have to pay for these things. That which is not reflected in price is deemed an externality.

Energy is essential to and promotes economic development. Note that we deliberately use the word development rather than growth, because we believe that the economy's role is to provide for the goods and services that improve our lives. Development is a better description of what we want. Without energy, there can be no manufacturing, no motorized transport, no computing or telecommunications, and much more limited agriculture.

---

[1] Some people have incomes of about a billion dollars per year, while others have incomes of hundreds of dollars per year – a factor of roughly 10 million difference, which is 7 orders of magnitude.

In the work that we refer to as the Energy Advantage (Martínez and Ebenhack 2008), we used the U.N.'s Human Development Index (HDI) as an indicator to assess the relationship between energy use and development. There is also a correlation, though, with Gross Domestic Product (GDP), which is a component of HDI, and which can be used to evaluate the direct financial benefits of energy consumption. Although the trend is not strong, it shows a definite correlation. There can be no doubt that energy contributes to wealth, even though it is exceedingly difficult to quantify.

In valuing energy systems, the financial costs and benefits must certainly be assessed. Not every type provides the same economic benefits. The highly dispatchable[2] and extremely energy-dense energy sources offer a great deal of value. Such fuels can provide energy as required, when required. This fact contributes to the favor enjoyed by the fossil fuels. All combustion fuels, though, have some of this benefit. Even the highly problematic use of raw biofuels, primarily firewood, provides for energy on demand. Charcoal draws higher prices, not only because it burns more cleanly than raw firewood, but because it is easier to transport and to control the combustion.

Economic Calculations and Factors

Economic value is one of the important aspects of energy value and, doubtless, the easiest to quantify. Indeed, efforts to quantify other values, such as cost benefit analysis, monetize other kinds of costs and benefits in order to be able to compare everything on a common basis. The quantitative analyses of energy economic value, then, are crucial to understand. As such, there are some important factors that we seek to calculate. In particular:

- Money is highly fungible: It provides a common medium by which almost any good or service can be exchanged.
- Economic calculations have appeal because of the potential to evaluate different possibilities on common footing.

---

[2] Recall that dispatchability is the quality of being able to be stored, transported and used when and where needed.

- Economic calculations are very sensitive to the time frame during which expenses are incurred and revenues received.

Economic Factors

Capital refers to the funds available to invest. The investor has choices to make and normally seeks to invest funds to offer the greatest advantage. That may be a direct financial advantage, or it may be something else the investor values. For some Americans in particular, this means luxurious cars and homes. Certainly, cars are not likely to return economic advantage, but the comfort, perceived safety, or status may be the values that affluent consumers seek. To be clear, like energy, money is not an end in its own right. Money and energy have in common that they facilitate acquiring the goods and services that we ultimately value.

The economic advantage of investments is commonly assed by rate of return (ROR), return on investment (ROI), and net present value (NPV). ROR refers to how quickly an investment is paid back—having much of the character of interest. If you make a loan, with terms of being paid only interest, then the interest would be exactly the ROR of making that loan investment. ROI refers to the overall profit made. If it takes a long time to make your investment back, then the issue of time discounting comes into play. Money is more valuable to you today than it will be in the future. There are other investments you cannot make with money that you are waiting to earn back from a previous investment. This challenge is referred to as an opportunity cost. You sacrifice opportunities to do other things with money while it is tied up in an investment.

A discount rate is applied in calculating earnings as time goes by, effectively reducing the value of future earnings by some percentage rate. This becomes critically important in evaluating the anticipated value of investments that will take a very long time to pay out. In essence, no matter how valuable something becomes in the distant future, it is not worth much of an investment today.

The role of the NPV concept is to account for time discounting. All of the future earnings are discounted, such that profits reflect the value in current money, discounting for the loss of value over time. It reflects what the future expenditures and revenues of a project as if they all occurred in the current year.

A related concept is risk discounting. Many projects have very limited likelihood of success. Therefore, when assessing the expected profits in uncertainty, the profits should be discounted for the perceived risk of failure. A significant example would be drilling exploratory wells in unexplored provinces. The likelihood of commercial success in these "rank wildcat" exploratory wells is typically under 10 percent. Most data commonly cited are for "exploratory wells" collectively.

In the United States, a PetroStrategies, Inc. report shows the success rates ranging between 50 and 70 percent (PetroStrategies 2015), but these numbers are for the United States (a very mature province) and includes "step-out" exploration, more than "rank wildcatting." Consider drilling in a totally unexplored province, looking for a possible discovery that is believed to have potential of 10 billion barrels of potentially recoverable oil. That much oil would be worth nearly a trillion dollars at prevailing prices! But, since the odds of success are only 10 percent, the average expected value of the discovery drops to US $80 billion. (The expected NPV drops even further when the many years it would take to produce all that oil are considered.)

Risk discounting should likewise be applied to unproven technologies. Controlled fusion is a prime example. It may ultimately have extraordinary value. At the time of this writing, though, the technology has not succeeded in yielding a sustained net energy gain. The risk remains that it will never be a commercial success. Therefore, the huge potential value of this new technology (and the tremendous resource that it taps) should be discounted for the risk of failure, as well as for the declining value over time.

Herein lies some of the reason why governments should support research and development of emerging technologies. Companies will be reluctant to make large investments, due to risk and time discounting, even if they believe that a new resource or technology will become enormously valuable. If controlled fusion is successful in another 50 years, beginning to produce a trillion dollars of profit, that would still barely justify an investment of $11 billion in 2014, if 10 percent/year is used as the time discounting factor. If the odds of success are 50 percent, the NPV drops further to US $5.5 billion, the maximum investment must be even further discounted for risk. It is not an attractive commercial investment now. Its promise must be pursued by governments. The question remains, even for governments, how much of the public's funds are worth pumping into uncertain technologies.

If, instead of the 10 percent discounting factor shown above, a 5 percent per year time discounting factor were applied, $110 billion investment is justified by NPV. The factor of 10 differences in the justifiable investment for a long anticipated profit also begs the question about how long investors are willing to wait. In general, investors expect to see some returns within about 5–10 years. Investment horizons of 50 years and more begin to be intergenerational. Questions can be raised about the appropriateness of using time discounting to account for time value in the intergenerational context.

In essence, no amount of profit is meaningful to a person approaching 60 if it is forestalled for 50 years (anything beyond 30 years becomes quite unlikely to be realized in a 60 year old's lifetime.) Most of us do have some concern for future generations, but that tends to be more of an emotional or conscientious value than a truly economic one. Time discounting economic benefits (or costs) may not be the appropriate way to assess value for the distant future though.

Indeed, time value is one of the challenges that climate change action faces. Even those of us who are persuaded by the evidence in support of anthropogenic climate change, cannot readily argue for the economic merits of investments to mitigate greenhouse gas emissions, if the severe effects will be seen in 100 years. Even at a 5 percent per year discount rate, the costs incurred in 100 years will be discounted by a factor of 131.

Applying the discount factor also depends on an economic premise of substitutability (see Ayres 2007). The idea is that money is the universal medium, which allows us to substitute one resource for another. True enough, so long as there really is a substitute for the resource being sacrificed.

If a few low-altitude communities are flooded in 100 years by rising sea levels, it can be argued that the residents can be relocated to higher ground more economically than preventing the warming that causes the rise in sea levels. The productivity of those coastal areas can be replaced as well. The sticking point is that some limits of substitutability exist. Will it be population pressure for habitable land if the population continues to grow? (At a very modest 1 percent per year growth, the world population will have increased by 270 percent in 100 years, resulting in some 19 billion people in the early 2200s.)

Will the limiting factor be food production, which may come into conflict with land for settling and re-settling such enormous populations? The food factor would be further exacerbated if we pursue biofuels from dedicated crops like corn or soy. Whatever it turns out to be, at some point, in a finite world, something will limit the ability to substitute for lost resources. All of these substitutions will depend on sufficient energy.

The noncombustion sources, especially wind and solar, have been identified as holding great promise for providing cleaner, more sustainable energy. The challenge that the popular "renewable" sources face is largely direct economic value. The advocates of renewable energy often claim that they are on the brink of outcompeting the fossil fuel stalwarts. This claim will certainly be borne out in sufficient time, when the fossil fuels are no longer able to keep pace with growing global demand. It may even be true sooner, as more external costs are internalized by taxes and regulation, but the case for outcompeting the fossil fuels soon is quite shaky, primarily due to the easy manipulation of facts to reach that conclusion.

For instance, Vanek et al. (2012) provide an example calculation of the economics of a photovoltaic (PV) system that shows a payout period of about 19 years for a hypothetical 2.2-kW PV system to be installed on a rooftop in Ithaca, New York. On the one hand, Ithaca is not an ideal location, being at a moderately high latitude and in a fairly humid, rainy climate. Hence, the economics could be better in some locations. However, a rather high price of electricity was assumed (13 cents/kW h). Most critically, the analysis counted credits and rebates of 55 percent of the total cost. That means that the full cost of the system would have a 38-year payout. That equates to a 2.6 percent rate of return—and a payout period that may well exceed the working life of the PV array.

This makes for a very marginal investment. Again, better locations might offer better economics. Any significantly higher electricity costs or limited availability of grid-based electricity would shift the analyses in favor of solar. In much of the Developing World, energy is of limited availability and/or very expensive; thus, solar PV can be an economically sound alternative today. Ongoing research is likely to lead to continued improvement that makes PV cells more efficient and/or less costly. In the meantime, though, PV provides serious economic competition to the established fossil fuel systems only if heavily subsidized.

It is popular to assert that the key challenge is to make the cells more efficient, but efficiency itself is not truly the key to an economically successful system. If solar cells were made twice as efficient at three times the cost, it would yield a less economically attractive system, not more. Greater efficiency is a way to increase the net useful energy, which can improve the economics, if the gain exceeds the cost to achieve such efficiency gains.

Recall that the typical automobile achieves about the same overall efficiency as solar cells. The low efficiency of internal combustion engines in individual vehicles has not restricted their use. Nor does comparable efficiency limit the use of solar cells, tapping a free energy source. The reason that efficiency seems more important for solar is that the energy source is so diffuse. It is hard to generate large amounts of energy, even though the resource base is enormous, because there is so little at any one point.

Another perspective on the importance of efficiency is that there is so much room to improve efficiency. There have been claims of 60 percent efficiency in some new photovoltaic cell research using novel materials (see Richard 2013). The cost of manufacture for current PV cells is already about equal to the cost of installation, which means that cutting the manufacturing cost in half will only reduce the installed system cost by 25 percent. However, if the efficiency doubles, then each panel will have twice the capacity and the same gain will be achieved as cutting the cost in half. It really is a balance of cost per watt hour produced. If that cost goes down significantly or if the price of the fossil fuel mainstays were to increase enough, then solar PV could be competitive without targeted subsidies.

The monetary cost of useful energy is what concerns consumers. We think that it seems ironic, at first glance, that people so much want their energy to be cheap, although it is of such critical value. At least casual economic thinking assumes that willingness to pay reflects the real value of anything. However, people certainly do not want to have to pay a lot for what is most essential. We certainly do not want to have to buy air to breathe. We do not want to pay a lot of money for water or even food. Energy seems to fall in that same category. We need a great deal of energy, so we do not want to pay much for energy.

Some people, including us writing this book, make deliberate choices to invest in alternative energy and efficiency, regardless of the fact that these investments do not necessarily provide economic advantage. Rather, these consumers choose to place value on aforementioned externalities. Only people with reasonable access to disposable income can really make such choices, whereas people living at the margins cannot.

Indeed, people at the margins may not be able to afford even energy investments that do have good rates of return. They must be able to come up with the capital. In a choice between food for one's family and investment capital for future profits, the answer is clear. Economics will not solve the problems of the world's poorest people.

Prices

The price consumers pay for energy is the most conspicuous economic cost of energy. It is frequently discussed and certainly cannot be ignored in valuing energy systems. One of the most popular topics of discussion about the problems of petroleum is the "high cost." We would challenge the notion of "sky high" prices for petroleum fuels. If that were true, would people really buy the least efficient cars? Would they buy homes distant from their work? Would we be eating foods from all over the world?

The price of energy has been—and continues to be—low. Those low prices have been instrumental to the growth of western economies. Prices can be shown to be low by invoking the basic time discounting principles of economic analyses discussed earlier in this section. When adjusting for time value, the price of oil and of gasoline has remained low, relative to the rest of the economy. Consider Figure 4.1, which shows the price of crude oil evaluated in constant dollars, adjusted for inflation showing a relatively constant value, except for a couple of large spikes.

Oil prices have been on a roller coaster ride over the past few years, at the same time that some claim that we have already passed the global peak in oil production. Others claim that we don't ever need to worry about running out of oil. The price spike in the first half of the current decade seemed to support the pessimists, while the current price collapse seems to support the optimists. What's really happening?

First some perspective: a portion of the price spike was a function of the declining value of the dollar. It appears that one-fourth to one-third of the spike can be explained by the dollar buying less of everything during that period. Of course, there is something of a feedback loop between oil price and the value of the dollar. Since the United States is so import-dependent, rising oil prices tend to have a negative impact on the value of the dollar. Still, all of this accounts for less than a third of the price fluctuation.

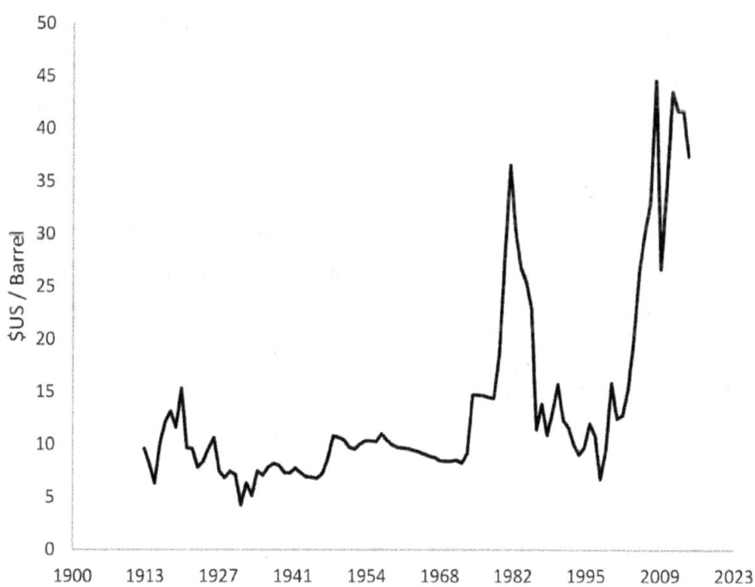

Figure 4.1 Historic crude oil prices, adjusted to constant 1984 US dollars (data were drawn from US DOE EIA).

Much of the rest of the fluctuation was a function of speculation. Some 2 million barrels of oil per day came off production during the height of the war with Iraq. Much of the lost production was being restored by the beginning of 2008, but prices were beginning to rise in response to lost production and anxieties were running relatively high about the possibility of war with Iran, which would imperil another couple million barrels of oil per day production. It is likely that these are some of the factors that caused speculators to bid oil prices up by nearly a factor of 2 in 1 year's time.

Another factor contributing to the speculation, though, was the proverbial little boy crying wolf, claiming that the world was already past global peak petroleum production. It seemed plausible and fueled the growing price speculation. Clearly though, the price spiked well above a stable point, in the absence of a physical shortage of petroleum supply.

Tremendous short-term growth from the shale plays in the United States may have increased production capacity sufficiently to influence the supply side of global petroleum markets, but the United States certainly has enough consumption share to influence the demand side. Consider that a 10 percent reduction in United States demand around 2008 equated to a 2.5 percent reduction in global demand. Couple that to reductions in Europe, and that likely was enough to push the pendulum of speculation in the other direction. Prices began to fall.

Also, American will seemed to be solidifying against new military action in Iran. The American, and then the global economies began to tumble at about the same time. This suppressed demand and created speculation that demand could fall even further. At about the same time, the dollar strengthened relative to other currencies, also tending to bring the oil price down.

One can argue that the oil price pendulum has swung disproportionately in the opposite direction. $50 per barrel is too cheap. What speculation gives, speculation can take away. Oil prices will rebound, perhaps overshooting a realistic stable point again.

This is not just confusing to consumers, but to companies as well. Imagine yourself as CEO of a major oil company. Should you be investing heavily in new exploration or alternative technologies? At $140 per barrel the answer would certainly be yes to new exploration, but at $50 per barrel it is no. Even when prices do rebound, how long would you want to see them stay up before you had confidence that they were not going to plummet again before your investments could bear fruit?

Energy alternatives, conservation, and the environment will benefit from relatively high, stable oil prices, but the free market is not giving us those. Only a shortage will signal the marketplace to raise prices on the fossil stalwarts. Only then will the prices rise to open the door for new energy systems to compete effectively.

Externalities

It can be argued that energy is so cheap because we do not pay for the negative externalities of consumption. External costs, such as in air and water pollution, are not always included in the price of a commodity. Taxes and regulatory policies serve to internalize externalities.

Probably relatively few legislators or industry advocates think of it in these terms, but that is the effect. Perhaps regulators and members of the public may think of regulations as punishing polluters. Industry officials probably think of the regulations as punitive as well, but perhaps view them as more arbitrary. Nevertheless, the effect of both taxes and regulatory prohibitions is to shift the costs of pollution and hazards into the marketplace, so that consumers ultimately pay for it.

This is another point that most policy makers probably do not want to highlight. It is certainly more appealing to voters to believe that the companies are paying for the taxes and regulations. Prices are determined, certainly, by the marketplace, not by the cost to the producer. But, if all the producers experience increased costs, there is a likelihood that the price will rise to the consumer. Indeed, we would suggest that the more apparent the cost is to the consumer, the more effective it is in terms of internalizing externalities.

The failure to internalize the total costs of the fuels we use to produce our goods and provide services is a way in which we fail to realize efficiency gains. We are all aware of a number of health, environmental, and social costs known to be connected to energy use from depletable, polluting fuels (e.g., increased health care costs due to pollution from coal use). The fossil fuels are all combustion sources, with all the impacts of the combustion products. While there is some tendency to conflate the renewability limits of fossil fuels with environmental impact, every resource system has its own externalities, which will become increasingly significant with increasing scale.

Consider two important factors:

- Every human activity produces unintended consequences, which can have unintended costs or benefits, which are not accounted for in pricing.
- Externalities can be internalized by regulation and taxation.

The idea here is that if these external, but linked costs are actually internalized into the cost of energy, then the increased price of energy will lead to incentivizing a host of improvements. It will lead to: improvements in efficiency; direct conservation, through reduced discretionary use of energy; new technologies; and shifting resources. We have seen conservation play out when oil prices rise suddenly or when the economy falters: petroleum consumption falls. Alternative energy sources gain some market share.

These all follow from higher prices. Increased price also drives greater production of the traditional fossil fuels, if the increased price goes to the producers rather than to taxes. All of this has previously led to a shifting balance of increased supply and decreased demand. Thus far, this shifting balance has always brought prices back down, setting back the march of efficiency and of new forms of energy.

Internalizing the external costs of energy could push prices up, without the likelihood of falling again. Sustained higher energy prices would give the alternatives a chance to gain market share. For instance, if a carbon tax is applied to mitigate greenhouse gas emissions, the cost of the dominant fossil fuel energy sources will rise. The noncombustion sources will benefit from the higher energy prices, without having to pay the tax.

There will be some pain associated with the higher prices, even though those prices will facilitate the expansion of alternative energy systems. Cheap, abundant energy has afforded the opportunity for amazing growth of industries, economies, and services: All based on massive and increasing consumption. The true costs of our consumption have been deferred, but must ultimately be paid. Yet we will still need many of the benefits of energy consumption.

We do caution readers that the focus on climate change can obscure many of the multitude of other external costs. There are numerous articles and books written choosing to focus exclusively on climate change, not merely as an important issue, but as *the* issue, choosing to define their system boundaries to include energy and climate. Certainly climate is a critical issue and the modeling of complex systems does require simplifying assumptions. Nevertheless, there are many more externalities that must be addressed—most of them are not as broad a threat as climate change, but may be more immediate.

Valuing energy systems adequately first requires acknowledgment of the problems. Then it requires assigning plausible, relative values to each. The task will be challenging. The effects are disparate and typically indirect. It will be difficult to find common ground for assigning costs to the wide range of externalities. Nevertheless, it will be essential to consider the full range of externalities for each energy system, to try to place them all on common footing. It will not suffice to regulate against or tax one set of external costs, while ignoring others.

Several factors play into the phenomenon of some problems remaining external to pricing mechanisms. In many cases, the costs to the environment or to society are too indirect to draw immediate attention. Especially in the case of environmental impacts, for centuries it was taken for granted that the environment was as big as the world and could absorb whatever wastes we poured into them.

As the scale of human activities and their associated waste streams grew, it did become rather quickly apparent that the local environments were finite and could be adversely impacted. Early responses were directed at moving pollution away from local communities. Smokestacks were made increasingly tall, so that the smoke would be dispersed by the wind at higher elevations.

Only within recent decades has the regional impact of the pollutants released been broadly recognized. Acid rain helped to catalyze this awareness. Primarily oxides of sulfur react with water vapor, resulting in the formation and deposition of sulfuric acid. Carbon dioxide and oxides of nitrogen can also contribute to this. The acidic compounds can be carried hundreds of miles downwind, crossing borders to acidify lakes and forest far removed from the point of origin. Regulatory policy helped to internalize costs, which led to more efficient, less polluting power plants. Acid rain dropped into the background of environmental issues. This is how it should be.

Monetization

A common effort to assess both positive and negative values is to monetize them. This may be uncomfortable for many people: Does assigning a monetary value cheapen disastrous consequences such as fatalities or environmental disasters? Nevertheless, some sort of common factor is necessary to enable comparisons. Money is the most universal, but monetary values are not simple to assign. Even once assigned, a major question is begged: Who pays for them?

For example, if a business owner's energy costs represent only 3–4 percent of their total costs to maintain a business space, they are less likely to invest in energy-efficient equipment and will allow society to pick up the costs of the mid- to long-term impacts of using relatively cheap energy in the short term.

Another challenge to addressing externalities is how to value them. How much is a little more air pollution worth (a negative value—in positive terms, it would be the value of cleaner air)? How much is it worth to reduce carbon emissions?

It is somewhat more straightforward to assess the costs of human health and safety impacts. Numerous precedents in the affluent nations are found to establish the value of a human life. Again, in spite of the moral repugnance of putting a price on human life, we do it all the time, in establishing life insurance values and in wrongful death suits. There is something inherently crass in placing monetary value on human lives, but it is a practical means to assign values, which can be used to compare with other kinds of costs.

Positive Externalities

Most of the discussion has focused on external costs: a typical perspective. Nevertheless, there can be unvalued benefits as well. When crude oil originally competed with whale oil, the price of crude oil had nothing to do with saving whales. Crude oil production, though, quickly dwarfed whale oil production, essentially ending the North Atlantic whaling industry. Alright, that point may seem like a bit of a stretch, but the positive externalities for all sorts of energy options need to be considered and valued appropriately, along with the negative externalities. A hydropower dam may provide external benefits of recreation and flood control, which are not included in the value of the electricity generated.

There can be argument about what to count in positive externalities. Reports by researchers at Argonne National Laboratories argued that the value of by-products of producing ethanol from corn should be counted as part of the net energy balance (Wang et al. 1999). Perhaps, however, does it really make sense to count the useful by-products of one energy source, and not another? This is quite extraordinary, considering the vast array of chemical feedstocks produced from petroleum.

In both cases, though, these are not externalities, they are products with value reflected in their own prices. We argue for making analyses as balanced as possible. The values of all by-products for all energy sources should probably be counted as credits toward energy production from each source. If we did not have synthetics from petroleum, there would still be a need for material for fabrics and for all the material needs served by plastics, for pharmaceutical feedstocks and so on. Without plastics, from what would we make all the stuff we use? There may be some reasonable answers, but the costs of the alternative products need to be considered as well.

Subsidies

One of the great sources of controversy regarding energy systems relates to subsidies. Are they essential to support the development of alternatives or do they distort the role the market should have in sorting the wheat from the chaff? Even if subsidies are useful for promoting alternatives, why do the well-established fossil and nuclear energy systems receive subsidies?

First, we need to clarify the meaning of subsidies. At first glance, the term invokes a connotation of direct payments from the government to companies or consumers. A large proportion of subsidies are actually tax deductions. Every person and every business is eligible for some forms of deductions. It seems worthwhile to separate subsidies that are given explicitly to support an energy system. Of course, direct payments would still apply. We prefer to call those "direct subsidies."

Tax deductions and credits targeting specific energy sources or technologies could count as direct subsidy as well. The key point here is that we believe that people are often misled by the statements about excessive subsidies, when, in fact, a large portion of the subsidy is a tax credit that all kinds of businesses or individuals could receive. We think that it would clarify the debate considerably to distinguish between general tax breaks and direct subsidies that target specific activities and tax breaks. Most ordinary tax deductions or credits are a function of total activity or revenue, making it seem that big companies are the most heavily subsidized.

Walmart gets far more tax breaks than a mom and pop hardware store, because they do so much more business and have a larger tax base than a small company. Similarly, both the tax payments and the tax breaks attributable to the big fossil fuel companies totally eclipse those from solar and wind companies, which are producing a tiny fraction as much energy, and thus a tiny fraction as much revenue.

The question is seldom asked in discussions of subsidies: what is the purpose of issuing them? Consider the following points:

- Subsidies refer to transfers of public funds to private entities.
- Subsidies may be direct payments or rebates, but they may also be tax breaks.
- They may include breaks that are not specific to the fuel or industry (they may be available to a wide range of businesses or energy systems).
- Some cash flows that are counted as subsidies to energy companies are doubtful.

Thus, subsidies that promote transitions to more beneficial and less hazardous or polluting resources add value, based on the extent that they promote the transition. In the Developed World, subsidies for solar and wind are often direct and are generally intended to promote market penetration of those resources. That has value in supporting industry, communications, transportation, education, and health services: all of the things that contribute to development and quality of life.

Direct Subsidies

Direct subsidies may be appropriate in some cases. Those that support conservation measures or purchases of specific kinds of energy alternatives may have value in promoting transitions to more sustainable energy systems. It can be argued (and we would agree) that the marketplace cannot be relied on to solve a true energy shortage. Since essential products, like energy, really must be available in surplus—and so many different companies and countries control large parts of the energy supply chain—it is unlikely that the marketplace will have a warning signal in time to respond effectively. Since there are reasons to favor some energy systems that are more sustainable, there is value in carefully selecting the most promising new energy systems to support the energy supply chain.

A controversial example in the United States was direct government support to a particular solar PV manufacturer. Solyndra received very large government guaranteed loans as part of the 2008 stimulus plan, along with a number of other emerging alternative energy companies. In 2011 they filed bankruptcy, defaulting on $535 million in debt to the U.S. government. Controversy is understandable. That's a large default from a single company, but Solyndra was not the only loss among the alternative energy loans from the government as part of the overall economic stimulus package (Washington Post 2014).

The emerging technologies, though, do need support if they are to progress sufficiently to take over noticeable market share from the fossil fuels by midcentury. Whether bets are made with investor capital or government funds, many of the investments in new energy systems will fail. Until the economics favor alternative energy, these investments are unlikely from private investors. Subsidies may be one of the few ways to prepare for the coming transition and provide support for new energy systems to build bases from which they can grow sufficiently as they are needed.

On the other side of the ideological spectrum, controversy surrounds subsidies to fossil fuels. The question is raised, why highly profitable companies should receive subsidies. They do receive subsidies. In some Developing Nations, fossil fuel use is subsidized directly to promote development. By establishing extremely low prices, governments can encourage the economic activity that goes with increased energy use and it can support transitions from more limited and harmful energy sources, such as raw firewood.

Promoting transitions away from raw firewood probably adds value, which may be worth governmental support. It leads to drastically reduced incidence of respiratory disease in women and children. It reduces pressure on the forests. It provides more useful and versatile energy. So, if a Developing Nation subsidizes natural gas, in order to promote transitions from firewood, charcoal and dung, it is probably beneficial.

Indeed, development activities depend on energy consumption. Thus it can be argued that subsidies that support consumption may support starting industry and other businesses that need energy. On the other hand, of course, if the subsidy supports increased consumption, it can be counter-productive.

It is essential, though, not to evaluate the future competitive potential of a resource (like solar and wind), based on the economics that are skewed by direct subsidies created to permit initial market penetration. Those subsidies will (and no doubt should be) removed once the emerging energy system is able to compete on its own. While the subsidy remains in place, the alternative energy system is not competing effectively. Saying that solar and wind are growing rapidly, in light of federal subsidy support is one thing. It is altogether different to claim that their rapid growth means that the mainstay energy systems are losing competitively.

Some subsidies are intended to support a product or activity, while others are simply a part of complex regulatory policies. Many things we call subsidies are neither direct, nor intended to promote the use of any given system. Indeed, many tax deductions that are broadly available to any industry are counted as "subsidies" because they reflect money that remains in the private sector that could go into the public sector.

So, as an individual, do you consider yourself to be subsidized by the government, when you take a standard tax deduction? Money that a business gets to keep, rather than going into government coffers, is counted as a subsidy. This can be misleading. In general, we will refer to companies or industries "benefitting" from subsidies rather than "receiving" them to mitigate against the impression that these are payments from the government to the companies.

Fossil Fuel Subsidies

In the past few years alone, there has been considerable scrutiny over what many categorize as fossil fuel subsidies, especially when compared with subsidies to (presumably) competing renewable energy systems. Numerous reports have been generated to highlight the perceived imbalance, but one in particular, from Adeyeye (2009) of the Environmental Law Institute (ELI) caught our attention as a noble attempt to sort out what these subsidies actually are.

The 2009 report itemizes and characterizes every subsidy they consider attributable to fossil fuel industries, which we depict in Figure 4.2.

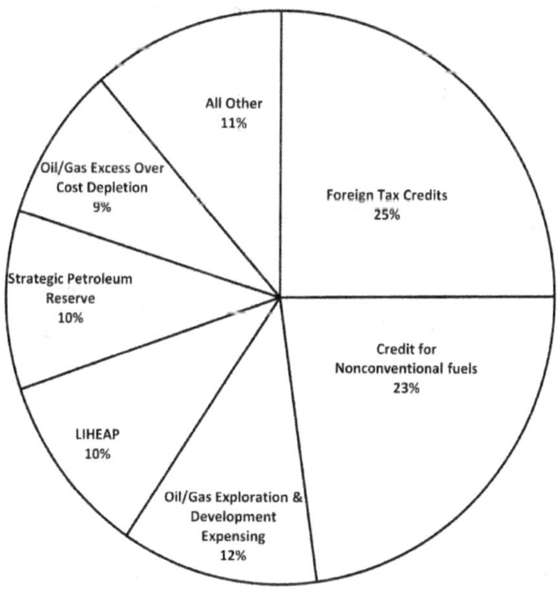

***Figure 4.2** Fossil fuel subsidies as characterized by the Environmental Law Institute (Adeyeye 2009).*

Based on these categories that they attribute to subsidizing fossil fuel energy, the ELI suggests that American fossil industries benefited from a total of $72 billion over the study period, while the much smaller renewable energy industries benefitted from a total of $29 billion over the same time period.

It is clear that the purpose of this report is to shine a spotlight on the disparity in funding between the fossil stalwarts and the renewable competitors. We do commend this report for being thorough in accounting and transparent in discussion of their logic for their categorizations, but we do question some of these categorizations used to get to $72 billion. Let us explore the categories to clarify the subsidies.

Tax Credits to Fossil Fuels

It can be seen that the largest single tax break in Figure 4.2 is the foreign tax credit. The ELI report clarifies that this is a credit available to all U.S. citizens and corporations doing some business abroad. It is intended to avoid being taxed by two countries on the same income. The controversy stems from the fact that some of what is counted as foreign taxes may actually be royalty payments to foreign governments.

The ELI report notes that, in 1995, the U.S. State Department suggested redefining some royalty payments as taxes. Of course royalty payments would still be business expense deductions from earnings, but this provision allows them to be subtracted directly from the company's U.S. tax bill, rather than from the gross earnings. That does represent a benefit to the companies—and decreased tax revenues to the U.S. government. The distinction between the treatment of royalties and taxes paid abroad would seem a worthy topic for debate in terms of energy costs and values.

But is every tax break truly what is commonly meant by subsidy? Since we, as individuals, do not view our own legal tax deductions as a government subsidy, why, then is it a subsidy when corporations receive tax deductions or credits that are peculiar to their circumstances?

Indeed, one of the small tax breaks identified as a subsidy is a credit for refining expenses to lower the level of sulfur in diesel. In the United States, relatively recent legislation sets stricter sulfur standards for diesel than for other fuels. We question whether it is useful to consider as a subsidy, a tax credit for expenses incurred specifically to comply with new regulations, which are stricter for that fuel than for others. We would suggest that a next step in evaluating subsidies might be to evaluate them in the context of what expense is being offset.

The tax credit question is somewhat problematic. Let us turn our attention to more direct subsidies. Indeed, the ELI study drew a distinction between tax breaks and direct payments. They elaborated the various categories of payments made from the government that reach fossil-based energy companies. Once again, their dissection of the categories is very informative.

Direct Payments to Fossil Fuel Companies

The ELI report also codifies what it considers direct payments from the U.S. government to fossil fuel companies. The four largest include the following:

- The Strategic Petroleum Reserve
- The Highway Trust Fund
- The Low Income Home Energy Assistance Program
- The Black Lung Disability Trust Fund

By far, the two largest payments are for the Strategic Petroleum Reserve and the Highway Trust Fund. It is difficult to argue that these two in particular are not subsidizing the American way of life. We run on oil and highways. However, what it is really doing is subsidizing the growth-based consumption model that we are accustomed to. Moreover, these two "payments" are really subsidizing transport systems, which we argue is independent of energy source. Would not the Highway Trust Fund still be in place if we were operating electric cars or even hydrogen-powered cars that run on electricity derived from the sun?

Also, try reflecting on the question: do we expect oil companies to fund the Strategic Petroleum Reserve? This opens a proverbial can of worms. Are we saying that we should restructure how energy companies operate in America? Even if the answer is yes to both questions and we created a national oil company akin to Statoil in Norway, we still would likely subsidize a strategic reserve of petroleum. It speaks more to a backbone infrastructure that renewable energy sources couldn't really compete with or even ameliorate anyway.

However, we would like to explore further a subsidy that the ELI report categorizes as direct payments, namely the Low Income Home Energy Assistance Program, or LIHEAP. The program does exactly what its name implies: giving direct payment to households to pay for heating oil or electricity. We question, though, that this should be counted as a subsidy to fossil fuel companies.

The ELI authors are thoughtful in targeting only the LIHEAP payments that are used to purchase energy. Nevertheless, we find it hard to extrapolate making energy purchases to prevent low-income households from losing heat and critical services as subsidizing the fossil fuel industries. Much of the energy purchased through this program would be oil for heating (especially in parts of Northern New England) and natural gas for heating and cooking.

First, the intent of the program is clearly not directed at supporting the fossil industries. It is meant to support essential energy needs of people who cannot afford it. If LIHEAP were eliminated and people were merely allowed, literally, to freeze to death in the dark, the oil or gas not sold to keep them alive this year would still be available to the companies to sell to more affluent consumers next year. The companies would still sell that energy, just a little later. Perhaps it would be viable to count the time value of accelerating these sales, but we still argue that this piece certainly does not match what people envision in hearing about fossil fuel industry subsidies.

It would be easier to argue that food stamps in the United States are a subsidy to farmers. Farm products are perishable, so the added food purchased by low income households does represent added sales for farmers (and grocers). On the other hand, the fossil fuels are not perishable: that which is not sold this year will be sold later.

We suggest that counting LIHEAP as a fossil fuel subsidy is very misleading, at least to people's common understanding of the concept.

LIHEAP represents about 9 percent of the total purported fossil subsidy and the largest piece of the "Direct Payment" pie.

Direct Payments from Petroleum Companies

Another factor to consider is the net flow of money. How large is an industry's total subsidy, relative to its total payments to the government? The fossil industries still pay taxes, in spite of the subsidies. They also pay royalties to the government for all of the value of oil, gas, or coal produced from government lands or waters.

In the United States, those federal royalties equate to one-sixth of the gross revenues. That amounts to approximately $10 billion per year for oil and gas. State governments also collect royalty payments. Contrary to popular rumors and in spite of the subsidies, the energy companies do pay taxes and lots of them. Of the 10 U.S. companies paying the largest taxes, the first, second, and sixth largest tax payers were oil companies, making payments in 2011, totaling $59 billion (USA Today 2013).

Compare this with the approximate $9 billion per year of subsidies attributed to all of the fossil fuel industry per year in about the same time frame. Indeed, there are also royalties paid to the government for the production of oil, gas, and coal from federal lands and waters. The government take is even higher.

Royalties

Many other national governments collect oil and gas royalties at even higher rates than the United States (US GAO 2007). Coal adds about another three quarters of a billion dollars of annual federal royalty payments (US ONRR 2014). These royalty payments represent large revenues from companies to governments. Typically, these are overriding royalties: a percentage paid from the gross value of the resource produced, before taxes or other expenses.

However, royalties only represent the tip of the iceberg. Companies also pay taxes on their own revenues. In addition to the $10 billion per year of royalty payments to the U.S. federal government, oil and gas companies paid another approximate $21 billion per year in other taxes and payments (API 2014).

The source of these data, the API, is an oil and gas industry organization. Naturally they have some bias. The environmental organization, Oceana, on the other hand, states that very little of the noted tax payments are made to the U.S. government: the foreign investment tax credits offsetting most of their U.S. tax bills (Oceana 2014). There is an important distinction.

The U.S. Energy Information Administration (EIA) report that Oceana used shows a combined roughly $9 billion in production, sales and use taxes paid by the petroleum industry. Another $3 billion is paid in "other taxes." That doesn't include payroll taxes or corporate income taxes. It is interesting that the EIA data cited in the Oceana report show a net refund to petroleum companies in 2009, but there were substantial net income tax payments to the U.S. government in other years. In fact, the preceding year reflected more than $23 billion in U.S. corporate income taxes (EIA 2011a).

Discriminating Subsidy Types

The EIA also prepared an interesting report on energy industry subsidies, in which they discriminate between types of subsidies. Direct payments, which we suspect most people imagine when they hear about subsidies to "big oil," are 0 both in 2007 and in 2010. Direct payments to collective Renewables rose from $110 million to about $4.2 billion. The tax relief for Renewable in 2010 reached $8 billion, about twice that provided to oil and gas (EIA 2011b).

The EIA report notes that they do not include all factors that could be considered to have some benefit to energy companies. They only include those that are provided by the U.S. federal government, with identifiable budget impact, targeted specifically to energy. They note that some major purported subsidies are excluded by only considering those explicitly targeted to energy. Some tax incentives that are available to all kinds of companies are not included. (See EIA 2011a for the full report.)

When the net cash flow is viewed on an energy-delivered basis, the picture may be surprising. The fossil fuel industries pay the U.S. government billions of dollars, net—after all of the subsidies. The solar and wind industries make far smaller net payments to the government.

This is not intrinsically inappropriate. The large, well-established industries can afford to pay much more into public coffers. The emerging technologies need some support to grow and to thrive.

Subsidies, though, should be analyzed, based on the goal that they support. Is it to support development of a new technology or a new industry? Is it to support the growth of other industry and business that depend on energy inputs? Is it to support household transitions away from highly unsustainable fuelwood dependence? If the goal is to facilitate the growth of a promising energy source or technology, it would seem logical that there should be plans for when and how to end that governmental support. If the goal is to support development, the subsidy should be evaluated based on how it meets that goal.

We also suggest that the language shift to reflect the difference between direct subsidies (e.g., grants) as opposed to the indirect subsidies of tax deductions. We believe that many Americans imagine the President sitting in the oval office, writing huge checks to huge corporations when they hear of large subsidies for fossil fuel companies. (Perhaps most of us realize it's not the President himself writing the checks, but the image of direct payments is still an issue that misdirects the debates.)

Tax deductions of all sorts, to all kinds of companies may need to be revisited, but that is a matter of how much of an entity's earnings they are allowed to keep. Then, comparing different sorts of entities (such as oil companies compared with solar power companies), it seems appropriate to consider the indirect subsidies as a fraction of their total contributions: both in terms of taxes paid and of energy produced. In those terms, oil and gas are much less subsidized (as fractions of the value they generate) than the renewable energy systems.

The enhanced subsidies (direct or indirect) for more sustainable energy systems do make sense. Perhaps it should be even larger, but that is a fundamentally different argument than one about "Why should highly profitable companies be getting money from the government?"

In effect, the answer to that question is that the fossil fuel companies aren't. The subsidies in question are not payments from governments to companies. They are payments not exacted from the companies. This still represents a benefit to the company. Indeed, the fossil fuel companies make huge payments to governments.

Some of the historic petroleum subsidies discussed were exceptions to the U.S. 1980 Windfall Profits Tax: a special tax that was applied strictly to oil and gas production (Hymel 2013). Deductions to taxes that only a specific industry pays count as subsidies to that industry. That seems problematic. Shouldn't those extra taxes be counted against the subsidy balance? We would suggest that a study of "net" subsidy might help to illuminate this convoluted debate.

A primary factor to be considered in assessing subsidies is the size of the subsidy, relative to the amount of energy produced. What share of the price that the customer sees is subsidy? The fossil fuel industries benefit from large subsidies, but produce tremendous amounts of energy. The U.S. fossil fuel industries (collectively) received a total of a little over $100 billion in subsidies (which were primarily tax deductions) during the period investigated. During that time, the alternatives received about one-third as much: $33 billion (Hymel 2013).

Many people argue that it is highly inappropriate to subsidize the lucrative fossil fuel industries, while the subsidies for emerging solar and wind are smaller, but their production is orders of magnitude smaller. So, on an energy-delivered basis, solar and wind receive about 100 times more subsidy than do the fossil fuels.

In some cases, as we've seen, companies seeking to carve out niches in the production of flow-limited energy systems receive direct, cash subsidies. These subsidies are speculative and many of them will fail to lead to important energy production. We believe, though, that these subsidies are important, largely because flow-limited energy production is not currently economically competitive. If we wait for the marketplace to respond to oil and gas shortages, there will be chaos.

Untested technologies for making photovoltaic panels will still fail, but those failures will require energy that is in short supply and time that is in even shorter supply. While energy remains cheap and abundant is the time to experiment with new technologies and for emerging companies to test their entrepreneurial wings. There may be time for some ideas to fail and other to gain some traction until they are able to stand on their own. Of course, they will never be totally without subsidies, since every company receives some tax breaks that are technically subsidies.

Subsidy Dollar per Unit Energy Delivered

With data derived from U.S. DOE sources (EIA 2015; AFDC 2015), we constructed a bar chart (Figure 4.3) as a preliminary attempt to characterize subsidies per unit of energy delivered (MWh basis) in the United States.

This paints a very different picture than simply looking at the gross subsidies provided to various industries. The net subsidy per unit of energy delivered is very small for conventional fossil fuels. Only "advanced coal," meant to minimize impacts such as climate change, receives significant subsidies. Corn ethanol has one of the largest subsidies on an energy-delivered basis. If we were take this evaluation one step further and base it on "net" energy produced, the American subsidy for corn ethanol would become astronomical.

Final Thoughts

Energy resources are priced in the marketplace by the dynamics of supply and demand. The demand for a given energy source is a function of values or merits of that source or its carrier. Those systems that hold energy very densely and are easily stored and transported tend to be the most highly sought. The growth of flow-based resources will be limited by this factor.

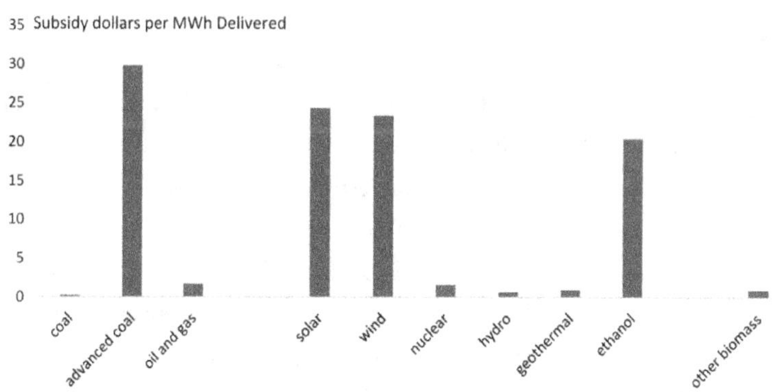

Figure 4.3 Energy subsidies issued per amounts of energy delivered by resource/fuel type (US$ per MWh delivered).

The price that we pay as consumers represents the most dramatic effect on the demand for energy. The term "externalities" refers to costs (or benefits) not reflected in the price to consumers. All energy systems have some external costs. Those that have achieved large-scale production tend to have the most readily recognized externalities. It makes sense that they would: both due to the scale of operations and time to observe the impacts produced.

Every human activity, though, has impacts. We should constantly be seeking to identify those impacts and assign costs to them. Those costs can be accounted for by regulations that either constrain or tax certain high-impact activities. These regulations will tend to be passed on to consumers, making the price of energy from given sources higher. The price is not directly set by costs to the producers. However, if the costs increase, the profitability decreases, which tends to decrease production. This, in turn, limits supply, which will tend to drive prices up.

Thus, we suggest that energy is undervalued in the marketplace. Since many of us do not pay much for energy, we do not treat it as a valued commodity. We should pay more. That will cause us to use it more judiciously. It will also facilitate the larger scale entry of new alternatives in the marketplace. Higher energy prices will be the single factor that will support environmental protection, reduced carbon emissions, and alternative energy.

References

Adeyeye, A. 2009. "Estimating U.S. Government subsidies to energy sources: 2002-2008." Environmental Law Institute. Accessed from: http://www.eli.org/research-report/estimating-us-government-subsidies-energy-sources-2002-2008

Alternative Fuels Data Center of US DOE (AFDC). 2014. "Alternative fuels data center: Fuel properties comparison." Accessed from: http://www.afdc.energy.gov/fuels/fuel_comparison_chart.pdf

American Petroleum Institute (API). 2014. "Oil & natural gas: Supporting the economy while paying our fair share." Accessed from: http://www.api.org/~/media/Files/Policy/Taxes/Oil-Natural-Gas-Industry-Pays-Its-Fair-Share-Taxes.pdf

Ayres, R. U. 2007. "On the practical limits to substitution." Ecological Economics, 61(1), 115–128.

Hymel, M. L. 2013. "Environmental Tax Incentives in the United States: Will Recent Market Incentives Reduce the U.S.'s Dependence on Oil." Arizona Legal Studies Discussion Paper No. 13-17.

Martínez, D. M., and Ebenhack, B. W. 2008. "Understanding the role of energy consumption in human development through the use of saturation phenomena." Journal of Energy Policy, Elsevier.

Oceana. 2014. "Myth vs. fact: Oil and gas subsidies." Accessed 2014, from: http://oceana.org/en/our-work/climate-energy/offshore-drilling/learn-act/oil-gas-subsidies-myth-vs-fact

PetroStrategies. 2015. "US drilling success rates." Accessed from: http://www.petrostrategies.org/Graphs/drilling_success_rates.htm

Richard, M. 2013. "Graphene could offer a 60 percent solar cell efficiency!" Treehugger. Accessed from: http://www.treehugger.com/solar-technology/graphene-could-offer-60-percent-solar-cell-efficiency.html

Tippee, R. 1993. Where's the Shortage? PennWell Press [Non-Technical Series].

US Energy Information Administration (EIA). 2011a. "Performance profiles of major energy producers 2009." Accessed from: http://www.eia.gov/finance/performanceprofiles/

US Energy Information Administration (EIA). 2011b. "Direct federal financial interventions and subsidies in energy in fiscal year 2010." Accessed from: http://www.eia.gov/finance/performanceprofiles/

US Energy Information Administration (EIA). 2015. "US crude oil first purchase prices (Dollars per Barrel)." Accessed from: http://tonto.eia.gov/dnav/pet/hist/LeafHandler.ashx?n=PET&s=F000000__3&f=A

US Government Accountability Office (US GAO). 2007. "Oil and gas royalties: A comparison of the share of revenue received from oil and gas production by the Federal Government and Other Resource Owners." Accessed from: http://www.gao.gov/products/GAO-07-676R

US Office of Natural Resources Royalties (US ONRR). 2014. "Reported revenues all land categories in all states and offshore regions for FY 2013." Accessed from: http://statistics.onrr.gov/ReportTool.aspx

USA Today. 2013. "Companies paying the most in income taxes." Accessed from: http://www.usatoday.com/story/money/personalfinance

/2013/03/17/companies-paying-highest-income-taxes/1991313/

Vanek, F. et al. 2012. Energy Systems Engineering: Evaluation and Implementation. New York: McGraw Hill. 978-0-07-178778-9.

Wang, M. et al. 1999. "Effects of fuel ethanol use on fuel-cycle energy and greenhouse gas emissions". Argonne National Laboratories, ANL/ESD-38.

Washington Post. 2014. "Special reports: Solyndra scandal." Accessed from: http://www.washingtonpost.com/politics/specialreports/solyndra-scandal/

CHAPTER 5

Resources and Environmental Costs

Energy and Environmental Impacts

All energy sources once converted and used by society will impact the environment. This is true whether the sources are of a depletable or nondepletable origin. However, the degree and scope of the impact will be a function of the type and amount of resource used. Thus, while there is no such thing as a clean energy source, the degree of cleanness can vary widely.

As we move from an era where energy use was guided solely by economic considerations to one that emphasizes environmental ones, we must catalog those impacts that most influence our decisions to use or reduce use of a particular source. We must also keep in mind the likelihood that the source that we are now trying to displace probably displaced a less clean resource and that a perceived cleaner resource may ultimately be dirtier than we thought it would be. Such is the case with energy transitions.

Typically, we equate environmental costs by a resource's tendency to pollute the natural environment, be it through air and water emissions, land alteration, and radioactivity, or through the use of additional resources and materials to complete the energy conversion. Emissions may come in the form of chemical, particle, or thermal releases into the environment, while land alterations may come in the form of physical movement of rock, vegetation, and soil.

Each energy resource presents its own impact, though many share common impacts, particularly fossil and biomass fuels for point source air emissions and renewables for land use needs. In this century, a particular emphasis has been placed on carbon dioxide emissions and water usage, which we consider as special cases. An overview of impacts is presented in more detail below, with an emphasis on commonly monitored and regulated point source emissions.

Coal Energy

Though it served as a viable and cleaner alternative to wood in a different time, coal now is the proverbial whipping boy of dirty energy in the Developed World.[1] Its origin is plant-based and it forms from compression over millions of years. Its structure consists of chemical compounds, comprising carbon, hydrogen, oxygen, sulfur, and nitrogen, with trace amounts of other elements, such as mercury.

Coal has a long history of use, initially for heating, but now is used predominantly in electricity production and in steel manufacturing. Thus, environmental impacts of coal use come from three main areas: (1) land via mining, (2) air via combustion, and (3) water via thermal pollution.

Land Impacts

As a solid, coal is typically obtained through mining, and the two most utilized methods for producing coal are underground mining and surface mining. Although we typically associate mining with images of workers traveling deep into Earth's crust via elevator to extract coal, the fact is, due to technological advances, surface mining has become the preferred method of extraction and represents 67 percent of all coal productive capacity in the United States (EIA 2015).

The most conspicuous surface mines are also commonly known as "strip" mines. That name deliberately conveys a very negative image, because prior to the Surface Reclamation Act of 1977, the impacts could be truly egregious. The enormous pits were sometimes abandoned with little or no reclamation of the original land, and were simply filled with water to make small, artificial lakes.

---

[1] As we have seen, the effects of firewood dependence are worse than coal's impacts, so coal can be a superior alternative to fuelwood in the Developing World even now.

Coal exploitation tends to incur longer lasting impacts on the land. Indeed, surface mining radically alters the local topography. At the very beginning of the operations, all of the native vegetation is stripped away, along with all of the topsoil. In addition to displacing habitat, this procedure radically alters the flow of water on the surface. The stripping process has often led to severe erosion, which could also carry minerals from the rocks exposed by the mining operations into streams, soils, and into underground aquifers.

Erosion was often appalling, and the severity of the problem led to the Surface Reclamation Act, which internalized the costs of mining by compelling companies to change operations dramatically, to restore the topography to something like its original contours, to replace the excavated rocks at the levels from which they originated, and to replant vegetation, whose roots mitigate erosion.

If one tours reclaimed surface mine regions, the type of vegetation may give a clue to how long ago the mining ended and the reclamation began. Small evergreens indicate recent activity. Large evergreens, of course, would indicate more maturity. Hardwoods are generally slower growing, so if mature evergreens are giving way to some hardwoods, the reclamation has been underway for a few decades. Though coal companies are required to return the area to an acceptable contour, using the preexisting topsoil, the act of mining produces a long-term change to the area that disrupts the local environment.

Mountain top removal is another, highly controversial variant on the strip mining approach in parts of the United States. Instead of removing relatively flat layers of overburden, the overburden is the top of a mountain. It is stripped off and the tailings are deposited in "valley fills." The United States Environmental Protection Agency (EPA 2013) reports some risks to water and to species. Although restoring grades is a typical part of normal surface mine reclamation, it does not appear to be a part of mountaintop removal. That is not to say that reclamation is not required in mountaintop removal, but that elevations would be extremely difficult to restore. A great deal of coal may be available to extract in such mountains, but the cost to the land is certainly larger than for many energy systems.

Air Impacts from Combustion

Combustion is a fairly straight forward process and only requires a fuel, oxygen, and a catalyst, such as a spark, to initiate the reaction. Thus, in a perfect world, combustion of a carbon-based fuel should produce only heat, water, and carbon dioxide. In reality though, impurities, including the need to use air instead of pure oxygen, result in a number of unwanted and harmful side reactants.

Waste streams from combustion almost inevitably end up in the atmosphere, which, in turn, may precipitate back down into water and land systems. The waste products that have significant health hazards are classified and regulated as pollutants. The classic pollutant categories include toxic gases, particulates, and compounds that undergo chemical reactions, which can form acids and other dangerous compounds. Each of them has direct adverse effects on human health and on ecosystems.

For coal, the EPA classifies the following as air pollutants:

- carbon monoxide, CO
- sulfur dioxide, SO_2
- nitrogen oxides, NO_x
- volatile organic compounds (VOCs)
- particulate matter (PM)
- mercury and other low-concentration air toxics
- carbon dioxide, CO_2

Carbon Monoxide

Carbon monoxide, or CO, is an odorless gas that is poisonous even at low concentrations. It is produced by the incomplete combustion of coal (indeed, all carbon-based fuels). CO can be generated not only by direct burning, but also indirectly in internal combustion engines. The mechanism for CO poisoning is quite efficient, as it binds with iron in hemoglobin much more preferentially than does oxygen. This results in decreased oxygen delivery throughout the body, which can result in death by asphyxiation at high enough concentrations.

Sulfur and Nitrogen Oxides

Sulfur oxides, particularly sulfur dioxide, or SO_2, is produced by burning of fuels that contain sulfur. The largest point sources of emission are coal-fired power plants, industrial facilities, and automobiles. SO_2 impacts the human respiratory system directly, but it also reacts with other atmospheric chemicals to form sulfate particles, or acids, which precipitate as acid rain.

Nitrogen oxides, or NO_x, are produced because of the presence of nitrogen in air. These oxides, like in SO_2, are both primary pollutants and also reactants in the formation of ground-level ozone. This brownish gas can also react with VOCs to produce smog on hot days. And like SO_2, nitrogen oxides react with other chemicals to produce acid rain.

Indeed, acidification of lakes and soils was the most popularly discussed environmental issues of the 1980s; however, the concern has been largely displaced in popular discourse by climate change. Not only is this partly a result of the extremity of the focus on climate change, but it is also related to the issue being somewhat ameliorated by improved technology to remove sulfur from coals both before and after combustion. Displacement of coal by natural gas has also mitigated the acid deposition problem to a large degree.

Volatile Organic Compounds

VOCs are another important pollutant. Organic chemicals are used practically in all household products, and are used in the production of any consumer good that has plastic or paint in or on it. Likewise, all fuels, from wood to ethanol, release small concentrations of VOCs all the time. They are not only most problematic in enclosed spaces, but also as a co-reactant with NO_x to form ozone, as mentioned above.

Particulate Matter

PM is probably the most widely spread type of pollutant. It is a direct combustion product, partly a function of incombustible contaminants in the fuel. Smoke is most commonly associated with PM, but also can be a mixture of various particles and liquid drops very small in size

(Reist 1993). Particle size and shape (sometimes called morphology) are directly linked to how they impact health, and most environmental agencies concern themselves with particles that are smaller than 10 μm (microns) in size. Though coal is a major contributor of PM pollution, we note that wood burning, too, is a major source of PM worldwide and is of great concern, particularly in the Developing World.

Mercury

Mercury is a naturally occurring element and is ever present in coal. When burned, mercury is released into the atmosphere, and eventually settles into bodies of water and land. If it makes it into the food chain, it can be absorbed as methyl mercury, which is highly toxic. The 2005 National Emissions Inventory estimates that coal burning is responsible for 50 percent of all human-caused mercury emissions (EPA 2015).

Carbon Dioxide

Carbon dioxide, CO_2, is an abundant gas in the atmosphere and is important for maintaining moderate temperatures on the surface of the planet. Excess emissions in recent times have caused concern that atmospheric concentrations are damaging Earth's energy balance with the sun, and negatively affecting global climate. Using evidence from a number of sources, climate scientists believe that we have approached actionable thresholds for mitigation.

There is argument about whether to include CO_2 as a pollutant. The point of treating CO_2 as pollution in United States debates is that it be regulated by the EPA. Since CO_2 is a primary driver of potentially catastrophic climate change, it makes sense to find effective means to consider regulating it. Indeed, technical clarity and regulatory needs are different issues. That is, pollution has technical meaning, but it also has regulatory meaning. We suggest that there is nothing wrong with characterizing it as a pollutant for regulatory purposes. (We will discuss climate later on in this chapter.)

Other Air Impacts

In addition to the combustion processes, pollutants enter the atmosphere at other stages of energy systems. Surface and underground mining liberates ultrafine particles, which are hazardous for workers.

Thermal Pollution

Water is used both as a coolant and in generating steam in thermal power plants. Thus, power plants were traditionally built along rivers or lakes, so that there was an abundant supply of fresh water for generating steam and cooling the discharge. The water is not lost in either process, nor is it chemically contaminated, but it can be hot when it is discharged.

The majority of this discharged heat stays within the riverine system, which can greatly impact the aquatic environment. Some fish species that have adapted to survive in specific temperature ranges are likely to die off as a result of the thermal pollution from the power plant. Other species that require certain dissolved oxygen concentrations can also be impacted. Plants and insects are also impacted, which ultimately can result in substantial degradation of the local ecosystem.

This problem can be significantly reduced by thermal recovery techniques, especially with combined heat and power, also known as cogeneration technology. It should also be noted that thermal pollution of waterways is a function of the conversion process, not of the source, so it will be a factor for any energy system employing thermal power generation.[2]

Oil Energy

Reservoirs of oil were formed in geologic basins, under conditions where rock sediments are deposited in areas rich in aquatic life, in which the water was relatively static, and under reduced oxygen to promote slow organic decay. The scale and time frame of processing result in a massive resource that can be tapped for extraction. Once oil is extracted, it is transported to markets and refined to be utilized as transportation fuels, heating fuels, and for chemicals manufacturing. Thus, the primary

[2] We suggest that cogeneration technologies be applied at all facilities: It increases the overall efficiency of the energy system as well as mitigates excess pollution.

source of pollution is air emission due to combustion and to evaporation, as well as water emissions via oil spills.

Air Emissions

Similar to coal, when oil is burned, it releases a series of criteria pollutants. They include all of the pollutants listed for coal save mercury, but can additionally include releases of chemicals such as lead, benzene, and formaldehyde, to name a few. These chemicals are formed at every stage of processing (production, refining, and storage) and the best way to control noncombustion air emissions is by regulating how fuels are formulated and also by regulating the recovery of vapors, specifically from gasoline.

Also, as in coal, ultrafine particulate emissions are a hazard generated from combustion, particularly from diesel fuel. And, there is a growing concern over local particulate emissions from the enormous hydraulic fracturing activities in the United States. The huge quantities of sand being handled as proppants for the so-called frac (not frack) jobs have been seen to release enough ultrafine particles to be potentially hazardous to the workers. In telephone conversation with Dr. Karen Mulloy, she related that her team went to sites of fracturing jobs, looking for evidence of atmospheric chemical contamination from the frac chemicals. Rather than finding chemical contamination, they found ultrafine particle contamination at the site (Mulloy 2013). Once again, there are problems, but not those that are commonly perceived.

Oil Spills

Petroleum has a particularly conspicuous potential to contaminate water. When contamination occurs by the doing of humans, it is often reported on when the event is massive and affects marine life. The prevailing issue regarding oil spills is that when it does happen as a result of some catastrophic event, it releases massive amounts of oil at one time. This will cause severe short-term effects that can wreak havoc on local habitats and fishing economies at the time of event.

However, most people (including many scientists and policy makers) often misunderstand and/or misrepresent how most spills occur. Offshore drilling does pose a risk for catastrophic releases of oil (and gas) to the environment. On the whole, though, this is a relatively small contributor to oil contamination in the seas. Several studies by the US National Academy of Sciences have shown that drilling and production are among the smaller sources of oil pollution in the oceans. However, according to a recent report (Osborn 2011), it represents less than 6 percent of the total contamination: barely over half as much as municipal runoff (see Table 5.1), and one of the categories within "Consumption Activities" and one-ninth as much as natural seeps.

Within the category of Transporting Petroleum, oil contamination from tanker shipping outweighs that from pipelines by more than a factor of 8. Thus, we argue that pipeline debates should focus on the fact that pipelines are the safest and most efficient way to move product, wherever possible. Pipeline leaks account for about 4 percent of total oil shipping spills that contaminate the oceans. One may note that pipelines more often run onshore. Some of their spills contaminate soils and aquifers. The same can be said of other nontanker shipping methods though. Of course, pipelines must be properly designed and maintained. Many spills are related to aging or neglected pipeline systems. Those concerned with preserving the integrity of water systems are well advised to push for the best practical pipeline designs and for regulation to ensure proper inspection and maintenance schedules.

Land Impacts

Current discussions in the early years of the 21st century center largely on petroleum systems, especially the shale plays incorporating very large hydraulic fracturing processes. The discussion and distress around hydraulic fracturing (or "fracking" as the media incorrectly calls it) have drawn a great deal of attention to land use in drilling.

Table 5.1 Share of oil contamination from various sources.

Contamination Source	Percentage of Whole
Natural Seeps	46
Consumption Activities	37
Transporting Petroleum	12
Petroleum Extraction	3
Other	2

Source: Osborn 2011.

The issue is complicated in several ways. First, it is true that the drilling operations require clearing about 5 acres of land for a drilling pad. Several more acres might be cleared to make a "lease" road to the location. However, with directional drilling and fracturing, six to eight wells are often drilled from a single pad, draining perhaps 640 acres (one square mile) of gas reservoir. Conventional vertical well technology would require dozens of wells, each on their own smaller pad and each with their own lease roads. The surface disruption and traffic can last for several months, though, since so many wells with long horizontal legs are drilled from the pad.

Once the drilling is finished, the site is cleared and reclaimed. Vegetation is replanted. Generally, plantings are selected to be fast growing, and sometimes landowners are disappointed by the selection of plantings. It may be important to plant some fast-growing species to mitigate against erosion. (The landowner can, however, specify wishes in the lease agreement to include vegetation that replicates what was originally there.)

Hundreds of truckloads of water are brought in to carry out the large frac jobs, along with many truckloads of sand (to prop the fractures open) and large pump trucks. All of this creates a good deal of noise during the drilling operations, as well as substantial traffic. Thus, the lease roads themselves must be substantial. Some landowners find all of the commotion to be more than they thought they had bargained for, as do passersby who may be delayed, waiting for some of the large trucks to maneuver their way to the leases.

Nevertheless, all of this activity related to drilling and completing the well ends and much of the land use can be rather quickly reclaimed. The wells themselves occupy very little space, so once the drilling rig and all of the associated equipment leave, much of the land can be replanted and restored to other use rather quickly. Roads must be maintained to be able to service the wells, but they do not have to be as large as the original lease roads for drilling.

The traffic impact from fracing is a very real one, with its own complexities. At meetings in Washington County in Ohio, discussions touched on issues related to Road Use Maintenance Agreements, or RUMAs (CEAO 2012). Local governments commonly require the oil and gas companies to maintain and even upgrade the public roads to account for the wear and tear that the numerous heavy vehicles will have on the roads. In some cases, there is actually concern about the companies being overly zealous in their road improvements, potentially leaving the community with expensive roadways that could create a long-term expense burden after the company ceases to maintain the roads.

Many local governments are recognizing the importance and the subtleties of well-structured road maintenance agreements to accommodate the large vehicle traffic. Properly done, these agreements can serve effectively to internalize what would otherwise be an externality. The companies generally recognize that these agreements are essential parts of operating at such large scales in modern times, and are typically reported as very cooperative. More difficult to mitigate is the impact on traffic flow, which will be discussed in greater detail in the section on social impacts.

Natural Gas Energy

The same processes that create oil also create natural gas, and the costs of exploration and production are somewhat similar. What distinguishes natural gas from oil, and coal for that matter, is that it is a gaseous fuel at ambient conditions. Moreover, it is composed primarily of a rather simple chemical compound, methane. One carbon bound to four hydrogens. This combination makes natural gas the cleanest of the fossil fuels and one of the most prized resources across the world for electricity production, heating, and chemical manufacturing. The largest impact of using natural gas consumption is from combustion, from direct venting and flaring, and in processing (exploration, drilling, and production).

Impacts of Combustion on the Air

In all respects, natural gas burns more cleanly than coal and oil. To give perspective on the degree of cleanliness, let us consider Table 5.2, which presents the number of pounds of air pollutants released per billion BTU of energy consumed.

Table 5.2 Pounds of air pollutants produced per billion BTU of energy consumed.

Pollutant	Natural Gas	Oil	Coal
CO_2	117,000	164,000	208,000
CO	40	33	208
NO_x	92	448	457
SO_2	1	1,122	2592
PM	7	84	2744
Mercury	0	0.0007	0.016

Source: EIA 1998.

It should be abundantly clear that if given a choice between coal, oil, or gas, then gas would prevail as the desired energy source from a combustion profile decision. Of course, oil serves better utility for use in automobiles (though a switch to natural gas might not be a bad idea). For electricity production, though, switching to natural gas has many appealing benefits over coal (albeit perhaps short term if the resource is depleted too rapidly). Especially from the perspective of home heating, it seems a better proposition to use natural gas over oil or electricity, especially when the electricity is derived from coal. Obviously, emissions are still being generated, but again, thermodynamics requires us to incur some penalty for our energy use.

Impacts of Venting and Flaring on the Air

Releasing natural gas to the atmosphere by flaring (controlled burning and release) or venting (direct, unburned release) is a common practice in oil and gas production. This is primarily due to safety reasons. As a gaseous, combustible material, improper or faulty handing of natural gas can result in a catastrophic explosion, harming people, and damaging equipment. The practice of flaring or venting is typical when gas production occurs in association with oil production from oil reservoirs and there is insufficient infrastructure to collect, process, and transport gas to market. The lack of a commercial market necessitates venting or flaring to reduce fire and explosion risks (see OGP 2000).

Flaring and venting are both environmentally problematic and economically wasteful. The gas flares of the Niger Delta are common images that portray the utter waste of a valuable resource that could be provided to a local market for positive gains. But even in the shale plays of North Dakota, flaring has become commonplace (EIA 2014). Burning gas produces a combustion profile similar to that of Table 5.2, while venting has implications on climate forcing, since methane is a strong infrared absorber. It makes both environmental and economic sense to reduce these practices as much as possible.

Biomass Energy Costs

Wood

The transition from biomass as a primary energy source to modern energy has only happened very recently in the Developed World and for the most part, has yet to happen in much of the Developing World. Indeed, as recently as 2009, 87 percent of the global biomass energy supply still comes from firewood (IEA 2009). Moreover, according to Smith (2006), energy from traditional biomass fuel accounts for nearly one-tenth of all current human energy demand. And of that, wood-based fuels comprise about two-thirds of household use.

So, despite the preponderance of attention on modern biofuels such as ethanol and biodiesel, we still must look at direct burning of wood and wood products as a major source of environmental impact. Consider cooking with a fuel that has roughly the same combustion profile of coal, and doing that in your home, without proper ventilation. In much of the Developing World, exposure is highest in poor populations, especially women and young children as these are the ones most often present during cooking.

Small particle exposure, in particular, has severe deleterious effects on respiratory and reproductive systems. Particulate pollution in enclosed cooking spaces is the cause of widespread respiratory disease in millions of people in the Developing World, resulting in over 4 million premature deaths every year (WHO 2014). When large urban areas grew and industrialized on the pervasive use of solid fuels, it has been seen to lead to smog and prevalent respiratory disease.

Modern Biofuels

Even for modern fuels such as ethanol and biodiesel, combustion profiles go beyond just carbon dioxide and water, but also oxides, PM, and VOCs. And just as fossil fuels release carbon that was stored in plants millions of years ago, biofuel combustion releases stored carbon. In the case of agricultural feedstock biofuels, that carbon may only have been stored recently.

That is the basis of arguing for a "closed carbon cycle" for biofuels; they are only capable of releasing recently stored carbon when they are burned. Generally true. However, the global carbon cycle includes biotic growth, storing carbon. Converting naturally productive forests to agricultural production of biofuel crops releases carbon that had been stored for decades or centuries and disrupts the storage process. It may be true that the biofuels result in less carbon dioxide emission per unit of energy delivered, but it is not sufficient to offer simplistic blanket statements about a "closed carbon cycle."

The big impacts we need to consider for the modern biofuels are those on land and water systems. The world has finite arable land. Whereas that is not a constraint in North America at this time, it may become one and it already is in some parts of the world. Both ecosystems and carbon balance will benefit from replenishing old growth forests. Hence, just as we cannot ignore the limits to crude oil merely because we have not encountered them yet, we cannot afford to ignore the finitude of arable land. Diverting more land to producing crops for fuel should be seen as a high cost activity, and ultimately unsustainable.

The water usage is also a large factor that seems not to have received adequate attention thus far. We suggest, in particular, that the water demands for agricultural irrigation in places such as the United States plains states represent a serious impact. Water that is pumped from confined aquifers depletes a valuable resource. Runoff can also be contaminated by chemicals fertilizers, pesticides, and herbicides.

It should be noted that the discussion above about land, water, and carbon cycle relates to growing crops for biofuel. These concerns become moot whenever the biofuel is derived from a waste stream: sewage or agricultural and food processing wastes. If the source is a waste stream, then the cost of production is attributable to the primary product.

Only emissions from combustion and processing need to be considered. In general, waste stream biofuels represent a win–win scenario. (We discuss more about water impacts at the end of this chapter.)

Nuclear Energy

Nuclear fission reactions release an immense amount of energy, which under controlled conditions can provide an abundance of clean electricity. Especially when you couple no air pollution and no carbon dioxide emissions with an extremely high energy density, nuclear energy has few rivals. Unfortunately, safety concerns from radioactive waste, accidents, as well as fears of weapons proliferation seem to marginalize nuclear energy in recent times. Accidents and proliferation aside, the two primary impacts that exist from using nuclear energy are thermal pollution of rivers and lakes (which was described earlier in the coal impacts section), and radioactive waste, which by far, receives the most attention.

Due to the nature of the nuclear fission reaction, the primary waste by-products from nuclear reactors and fuel processing plants are radioactive. Within that general descriptor, there are wastes classified as high-level radioactive waste, such as spent reactor fuel from nuclear electric power plants, as well as low-level radioactive waste such as waste from fuel processing and reactor operations (see Karam 2001).

Spent reactor fuel, that is, uranium fuel that is no longer efficient at producing electricity, is both thermally hot and highly radioactive, due to the isotopes produced during the fission reaction. Although these isotopes will decay or disintegrate to harmless materials, the time frame over which radioactivity reaches safe levels for human exposure can be quite long. Indeed, some of these materials can remain radioactive and fatal for thousands of years. As such, the spent fuel can only be handled remotely and behind shielding.

All nuclear plants in the United States are required to store high-level waste in what are known as spent fuel pools. These are water-based storage units comprising reinforced concrete several feet thick, and lined with steel. The water level is about 40 feet deep, which serves to both shield the radiation and cool the spent rods. There are currently no facilities that are able to store nuclear waste long term (NRC 2015).

There has long been pervasive anxiety about the disposal of high-level radioactive wastes from nuclear power plants. Certainly, these concentrated radioactive materials pose serious threats to human safety and to the environment and must be disposed of properly. But what makes proper disposal? Since we know the half-lives of the radioisotopes, we are able to apply numbers to how long the waste should be contained to avoid a catastrophic release of radiation. Some of the half-lives are very long and are cited as reasons to be especially frightened of radioactive hazards. This misses the point, though, that the hazard is diminishing with time, which may not be true of many chemical hazards, such as heavy metal accumulations that can occur with coal-based energy.

Hydropower Energy

Hydropower has been the hallmark of renewable energy for several decades. However, the deliberate alteration of natural water flow processes on the surface of Earth to generate electricity does result in impacts that can be significant. This is especially true on the scale that humans have utilized hydropower in recent times. In particular, large dam and reservoir systems have generated a substantial amount of impact on the land, and on ecosystems. And due to the scale of the installations, it becomes very difficult and costly to decommission a project and demolish the structure.

Indeed, hydropower has fallen out of favor as a sustainable energy resource, in large part due to its disruption of natural river systems. Even though the water doesn't go anywhere and is not contaminated by chemicals, its form changes from flowing water to large bodies of standing water, which represents totally different ecosystems. Hence, while hydropower plants create large quantities of clean and reliable electricity, they can result in the following impacts:

- Obstruction of migration of fish to upstream spawning areas
- Alteration of natural water temperatures, chemistry, flow characteristics, and silt loads
- Coverage of important natural areas and agricultural lands
- Generation of greenhouse gases, particularly methane

One of the earliest concerns related to hydropower impacts on aquatic systems in the United States was that it would interfere with the migration of fish, like salmon. Hence, dams began to incorporate fish "ladders" though it turns out that's not a very good description. The ladders are actually sets of concrete steps at the side of the dam, with water flowing down over them. The fish can swim up, leading from one stair step to the next, as long as they can find the ladder. In fact, salmon may expend too much precious stored body energy, just swimming around, seeking the way forward (Helfman 2007). Whether migration is disrupted or not, the ecosystem is drastically changed and ecosystems do not respond immediately.

Wind and Solar Energy

The generation of electricity by wind turbines or by solar photovoltaic panels, at the point of their use, does not generate any combustion emissions, radioactivity, or other typical nonrenewable contaminations once installed. Indeed, large-scale installation of either energy system likely would result in a significant reduction of carbon dioxide emissions, especially when compared with coal energy. They also do not require the consumption of water to generate steam as in traditional fossil and nuclear power plants. Thus, it should not be surprising that growth in installed capacity for both energy resources has climbed rapidly and will continue to climb for the foreseeable future for their perceived environmental benefits.

However, despite the wonderful potential for cleaner energy production from wind and solar resources, it is prudent to recall from Chapter 1 Romer's three facts about the use of energy, namely (1) the energy we use has to come from somewhere; (2) the energy we use has to go somewhere after we use it; and (3) energy conversions have side effects that may be undesirable and that may not have been anticipated.

Thus, as we continue to champion the proliferation of solar and wind technologies to harness renewable and sustainable power, we must be made aware of environmental costs associated with the utilization of each resource, as well as the likelihood that as time passes, the list of impacts will increase.

Wind Impacts

As of this writing, the focus of direct impacts from wind turbines includes noise, electromagnetic interference, wildlife, and aesthetics.[3] Noise pollution and aesthetics in particular can be very difficult impacts to regulate. It is true that a wind turbine's spinning whoosh generates noise and that a wind farm can obstruct the view of a pristine location, but how to mitigate, or even recognize as actionable impacts, is still new territory.

For noise in particular, it is likely that the sound level generated from a wind farm (as measured on the decibel scale) is lower than that generated by passing automobiles or trucks. A difference is that the constancy of the whooshing sound compared with less regular traffic noise can in some instances severely impair the health of local residents. Another difference is that very low–frequency infrasonic noise can also be transmitted to nearby residents and can be potentially harmful.

However, just as not everyone is as sensitive to certain pollutants, not everyone is sensitive or susceptible to these new forms of potential pollution sources. Indeed, the small number of cases of "wind turbine syndrome" that get reported on in popular media (see Upton 2014) may simply represent new kinds of environmental sensitivities that are widely found with exposure to all types of man-made, synthetic materials. Although the sample size on these potential impacts remains small, these issues will likely become more prominent as wind farms proliferate in areas with higher population densities.

The aesthetics impact is even more difficult to assess. A complete ban on a technology or facility as the result of being unsightly is highly problematic. This speaks more to the psychology of public acceptance of a technology than it does to any real environmental impact, but it is obvious that it should not be discounted. This is especially true when wind farms are being sited in what many consider to be pristine environments.

---

[3] There are also a number of indirect impacts that are associated with wind energy, particularly those associated with embedded, or embodied energy. That is, each wind turbine that is created requires significant amounts of steel, coal, and petroleum resources in the form of fuels, electricity, and plastics that must be accounted for when assessing the total environmental impact of utilizing the wind resource.

Finally, it is undeniable that from a point source perspective wind requires large, open spaces. (If wind turbines are spaced too closely together, they interfere with each other.)

Direct Solar Impacts

Solar energy requires a great deal of space, because of the highly diffuse nature of the energy source being tapped. In the case of very large-scale applications, large tracts of land must be dedicated to the solar power plant. This is especially true for concentrating solar power plants, which can affect habitat. However, one of the great appeals to solar PV is the great potential for rooftop mounting, in which the solar system does not require any new space.

With respect to solar PV in particular, the primary concern has to do with hazardous chemicals used in the manufacturing process. These chemicals are virtually identical to those used in the semiconductor industry, and include mineral acids (hydrochloric acid, sulfuric acid, and nitric acid), as well as other chemicals such as hydrogen fluoride and acetone. Additionally, silicon dust can pose a potential risk to workers, and must be regulated.

A Word of Caution: Size and Scale Affect Impact

We feel a word of caution should also be placed on direct impacts. As more wind farms or more solar parks are constructed closer to residences, environmental issues may become more prominent and new impacts may be uncovered. One particularly interesting recent finding in Texas (which houses some of the largest wind farms in the world) is that night time temperatures in the local area have increased about one degree since construction of the power plants (Zhou et al. 2012). We know that even small degrees of thermal pollution in waters can impact ecosystems, so it is not unreasonable to believe that the same can happen on land at these sites.

Extensive solar PV manufacturing activity likewise incurs the potential to yield extensive pollution, especially with respect to thin film PV technology. Consider Kodak Park. Eastman Kodak has been a highly respected company, playing an integral role in the growth and memories of America and much of the world. They grew from a startup company to popularizing photography for everyone. It was a classic example of the growth of an industry to meet exploding global demand, perhaps an example for the future growth potential of solar photovoltaics.

Thin film processing, like that done at Kodak, caused the Kodak Park facility to be known as the biggest polluter in the state of New York (Niman 2003). It was the enormity of the scale of manufacturing that opened the door to the enormity of the pollution. Even if thin film solar photovoltaics reach an even larger scale of production than the peak of Kodak films, there may still not be catastrophic release of massive pollutants, but such large-scale operations require monstrous scales of chemicals and energy use, which opens the door to serious pollution.

On Climate and Impacts

In general, the popular discussions of climate refer to human impacts on the global climate system. In early discussions in the 1980s and 1990s, the issue was commonly referred to as "global warming," which promoted images of relatively continuous temperature increases. These images are inaccurate and prone to the erroneous notion that cold spells disprove the theory. Indeed, whereas the problem relates to increased absorption of infrared radiation and subsequent warming, the outcome is much better described as a destabilization, hence the more contemporary use of "climate change."

So what is the energy connection? Well, we know that burning fossil fuels does release billions of tons of carbon to the atmosphere as CO_2, which had been stored in Earth for millions of years; atmospheric CO_2 is increasing; and CO_2 is a "greenhouse gas" that absorbs infrared radiation, tending to warm Earth. Extensive scientific work has created complex climate models showing that this trend of continuous addition of CO_2 into the atmosphere has put us on a road to a destabilized global climate, with likely serious consequences.

Will it be as bad as the models suggest? Will it be worse? We can't know until the events unfold. The fact that the system is too complex, with too many variables for anyone to calculate direct outcomes is why models are used. However, the best science tells us that continuing down the current path could have disastrous consequences.

There can be no doubt that human activities are releasing tremendous amounts of CO_2 into the atmosphere, by burning organic materials that had been long stored in the crust. There is no doubt that Earth is experiencing increase in carbon dioxide concentrations in the atmosphere. There is no question that CO_2 absorbs infrared radiation, which will tend to warm the atmosphere. The questions revolve around the complicating factors: positive and negative feedback loops and other climate forcing phenomena.

Rather than engage in the open ended debate about the extent of climate change, let's take at face value the general scientific consensus that the phenomenon is real and significantly anthropogenically driven. Like wearing a seatbelt, one need not be convinced that an accident is imminent, but merely that the result of a potential negative outcome is sufficiently severe that it makes sense to try to take appropriate care. The risks are clearly unacceptable to gamble with anthropogenically induced climate change. Thus, we are well advised to take seriously a goal of mitigating it.

Analyzing Potential Climate Costs from Energy Choices

We can establish a baseline of relative climate forcing for each energy system, which can provide some clear winners and losers on this criterion. It is abundantly clear that the combustion fuels all have ongoing greenhouse gas emissions and that makes coal, oil, and gas marginal sources to utilize from the climate forcing perspective. The degree by which each of these fossil fuels impact the Earth-climate system does, however, vary and can be controlled by both technology and consumption choices. Of the three resources, we would argue that coal energy is the baseline to which oil and gas and the other energy systems be compared.

Biomass and Carbon Costs

As we have alluded to previously, some people refer to biomass as "carbon neutral" but we argue that this is specious. Biomass fuels emit a comparable amount of carbon dioxide per unit energy delivered as does coal and far more than do oil or gas. Biomass burning, in the form of open fires, has been estimated to account for as much as 18% of carbon dioxide emissions (Luoma 2010). This method of energy consumption, which provides the primary energy for half of humanity, also releases an abundance of particulate soot.

Most notably, biomass use is associated with deterioration of land systems from forests and grasslands (which naturally sequester significant amounts of carbon) to lower value croplands, or, in the case of fuelwood dependence, it can result in utter degradation to barren or desert land. Both of these transformations shift the balance toward increased atmospheric CO_2.

Additionally, the notion of a closed carbon cycle implies a prompt balance, which is patently untrue. Carbon is released quickly when the material is burned. It takes months to years—or even decades—for the growth of new biota to take up CO_2 from the atmosphere. Even for seasonal crops, like corn or soy, the crop is harvested, converted to fuel and burned, but it is months before the next crop of corn or soy can be grown. Demand goes on constantly, whereas crop production follows specific cycles. The new growth does take up CO_2 from the air.

But, if crops were not being grown on the land, what else would be happening with it? Would natural vegetation be growing there? Would the plants be storing carbon? If so, the growth of plant material cannot be attributed to the energy system, it is a natural process for the land. Therefore, it is very questionable to attribute the carbon uptake by new growth to the fuel crop.

The natural growth may be consumed by wildlife, and it may be part of old forest growth, which fixes carbon for several decades. Of that which dies and decays, some of the natural decaying biotic material will contribute to soil building. Some decay products will yield methane. The balance between methane emitted through decay processes in natural ecosystems compared with the CO_2 emitted to the atmosphere by combustion or land clearing is a complex one. Some decayed organic matter will actually be carried with runoff out to sea, to be buried with sediments and eventually become future fossil fuels.

Methane

While CO_2 is the main contributor to anthropogenic climate drivers, it is certainly not the only one. The next most prominent factor is methane. It is much less prevalent, in terms of concentration, but it is a much stronger greenhouse gas than CO_2. Methane has a greater ability to absorb infrared radiation, somewhat offset by the fact that it does eventually oxidize to CO_2. Accounting for these factors, each molecule of methane has roughly 20 times as much greenhouse effect as a molecule of CO_2; so methane is also important to evaluate.

A recent article in the New York Times cited a Cornell University professor as saying that methane emissions from natural gas productions are likely to make natural gas as bad as, or even worse than, coal for climate change (Ingraffea 2013). The concern he raises makes intuitive sense. Natural gas is mostly methane. Some of that does escape to the atmosphere through a variety of leaks. Methane is a strong infrared absorber.

The missing piece of the discussion is a comparison with competing sources, including coal itself. In fact, coal mining is responsible for about 10 percent of all annual methane emissions compared with 29 percent for oil and gas combined (EPA 2014). Oil and gas combined, then, produce 2.9 times as much methane as coal. But wait. Oil and gas combined also produce about 2.9 times as much energy as coal. The amount of methane produced per unit of energy delivered is about the same. So, just comparing methane emissions on an energy equivalent basis, the claim becomes problematic.

Even if the argument is made that neither coal nor natural gas should be used, but rather solar or wind, it still remains moot because natural gas and coal aren't even the largest sources of methane emissions. Indeed, most methane emissions come from landfill and agricultural sources. Combined, they represent 57 percent of all methane emissions. (It seems that we should be worrying more about how much beef we eat, since that is the source of most methane emissions in America.)

Of course, there's also the fact that coal emits far more CO_2 than natural gas on an energy equivalent basis. The CO_2 emissions account for seven and a half times more climate forcing than methane. Coal emits 52 percent more CO_2 than methane, with somewhat less overall energy production. This clearly overwhelms any additional climate forcing for natural gas versus coal.

Furthermore, natural gas is less energy intensive to produce, or to ship. Gas is also more efficient in producing useful work or end use energy than coal; so natural gas clearly does offer great potential for mitigating climate change if it substitutes for coal. The substitution can be one of the readiest and most effective ways to reduce greenhouse gas emissions—other than conservation.

Noncombustion Sources

Noncombustion sources are generally considered to have no climate forcing impact. This is not technically true. First, they all require construction and manufacturing. As of this time, most of the energy to build any energy system comes from fossil fuels, thus generating some greenhouse gas emissions. Consider estimates that it takes crystalline silicon solar cells about 4 years to generate as much energy as it takes to build manufacture the cells (NREL 2014). That means that it will take at least 4 years of energy production by solar cells before any of that can be considered truly carbon free. Wind and hydro power plants require large quantities of concrete, which also require large amounts of fossil fuel input to create. It takes time to produce energy to offset those fuel inputs.

On the other hand, a successful oil or gas well can repay its energy investment on the first day. Indeed, a massive effort to manufacture solar cells and wind turbines will undoubtedly yield an increase in greenhouse gasses in the first decade or two. There is no room for rampant optimism about quick and easy transitions to a low-carbon society.

Nuclear power can produce a great deal of energy quickly, but does require tremendous amounts of concrete and steel. It is the energy option that has the capacity to take a large market share rather quickly, but it, too, will have large climate forcing in the construction phase, which will take more than a decade for each facility.

Time Considerations

Regardless of the alternative energy source, time is working against climate mitigation. Energy will become less carbon intensive. Noncombustion fuels are making rapid gains, building on their currently tiny market shares. This growth will continue and fossil fuel production will ultimately begin to decline, but the growth of renewables will not displace fossil fuel demands—or their carbon emissions for some time.

Conservation is the only energy activity that can significantly reduce greenhouse warming in a short time frame. However, global net conservation is an unrealistic outcome, since the affluent want the advantages and luxuries afforded by energy use, but, at the same time, the world's least affluent need more energy in order to meet their most basic needs.

A limiting factor in societal responses to climate change is the value that energy provides in our lives and our economies. In an interview launching a new book about Russian gas relative to the 2014 Ukrainian crisis, the Chairman of the Oxford Institute for Energy Studies observed that climate change is very low on the agenda for the average Russian, although local environmental impacts are viewed as much more important (Stern 2014). Consider, then, how much more locally focused people's concerns are in less affluent nations. If immediate needs are not being met, it is exceedingly difficult to embrace larger issues.

Another limiting factor is the lack of immediacy. Those who wish to deny climate change will have a long time to do so before they are absolutely proven wrong. In early 2014, reports came out of studies finding that the inertia of global warming is such that a large portion of the Antarctic ice sheet is already irreversibly on its way to break off, raising sea levels disastrously. This is dire news, which has drawn media attention.

But, the reports suggest that this will happen in 200–500 years. Two to five centuries in the future? What else may happen in that time? How much confidence can be placed in such long-term projections? The findings do serve to emphasize how severe the implications of climate change are and how intractable the problem is. But, it is so far in the future, that it will be readily dismissed by the naysayers. Perhaps even worse for persuading people to change our behaviors is the conclusion that this disaster is already unavoidable. Hence, one may ask, why bother to change, if it's already too late?

On Water and Impacts

Some energy systems have very conspicuous water demands and impacts: Oil and gas receive a great deal of the attention for both the potential to contaminate water and the direct consumption of water. Mining activities, especially for coal, have a potential to contaminate subsurface aquifers. Agricultural biofuels require a great deal of fresh water, both for growing the crops and for processing the fuels. Large-scale hydropower systems dramatically alter the flow of rivers.

In this century, the threat of water shortages appears to loom heavily, and water is certainly one of the critical resources for the survival of humans, and all other forms of life on Earth. Anything that adversely impacts this resource incurs a serious cost. Essentially all human activities require some water, whose needs must be considered. Competing energy and water needs must be addressed and solutions in place to ensure the sustainability and vitality of natural systems for societal benefit.

Coal Mines

Coal mines can pollute waters both during their operation and for decades after they are abandoned. Waters drain from and through the coal mines, leeching out minerals, often becoming acidic. Coal mine drainage can lower the pH of streams from approximately neutral, which would be a pH of 7, to less than a pH of 3. This level is roughly comparable to the acidity of lemonade, enough to kill fish populations.

The Western Pennsylvania Coalition for Abandoned Mine Reclamation (WPCAMR) was founded just 4 years after the United States passed the Surface Mining Control and Reclamation Act of 1977. The organization works with industry, agencies, and citizens to improve the pollution caused by over a century of runoff from working and abandoned coal mines (WPCAMR 2013).

Of note is that some oil and gas companies are seeking to tap the truly toxic waters from Abandoned Mine Drainage (AMD) for use in frac waters. Oil- and gas-producing companies have begun to partner with environmental nonprofits in Pennsylvania that are struggling to remediate the runoff from hundreds of old coal mines. For the most part, these agencies are strapped for cash to handle the cleanup of so much water from so many mines. The oil and gas companies can bring the financial resources to bear, while helping to take hazardous mine water out of the environment and put it to use, helping to produce more energy, more cleanly.

Seneca Resources has already fractured more than 75 wells in Pennsylvania, with frac water composed of 75% reclaimed AMD water and 25% recycled water, extracting as much as half a million gallons of contaminated water per day. In order to proceed with this sort of apparent win–win partnership, the oil companies need some assurance from regulatory agencies. In particular, they need to know that they are not assuming liability for the already contaminated water in perpetuity by using some of it. The agencies have indicated interest in working with the companies (Rassenfoss 2013). This sort of effort may demonstrate the value that cooperation adds to energy system development.

Hydraulic Fracturing

Speaking of fracturing, and perhaps behind the motivation to use AMD, is the concern about water usage for the massive hydraulic fracturing processes employed to develop oil and, especially, gas production from tight, relatively impermeable, shales. Fracturing is essential in order to allow significant quantities of hydrocarbons to flow to the well-bores.

The concerns about water are two-fold. Perhaps the most difficult to address is the risk of frac fluids escaping from the productive formation to contaminate surface waters and shallow aquifers, while the less difficult concern relates to the amount of water that remains in shale formations after it is used to fracture the rocks.

It does matter from whence the water used is taken. As of this writing, frac fluids are commonly constituted from fresh water. If the water is drawn from lakes and streams in relatively fresh water rich areas, then the concern is less urgent if you consider that water is a by-product of the combustion process. To do that, we have to think about the overall mass balance of the combustion process.

Oil and gas are chains of carbon and hydrogen, with a little more than two hydrogens per carbon. The molecules react with oxygen to mostly yield: energy, carbon dioxide, and water. Therefore, water is released with every molecule of oil or gas burned. If we are doing our stoichiometry right, that means for a well producing a modest 1 million cubic feet of gas per day, we are producing up to 83,000 pounds, or 10,000 gallons of fresh water, most of which enters the atmosphere. Relative to the amount of water used in a frac job, there seems to be an acceptable trade-off.

Naturally, this does not mean that water use is not a concern. Putting water into the atmosphere does not guarantee that the water will return to the fresh water system. It enters the atmosphere perfectly fresh as vapor. However, some of the precipitation will fall over the oceans, mixing with the salt water there. Nevertheless, the overall oil and gas system no doubt results in an increase of fresh water on the Earth's surface.

Fresh water released to the atmosphere will ultimately return to the Earth's surface as rainfall, which is likely to find its way to streams. Of course, it matters, as well, where the produced gas is consumed. If it is shipped long distances, the rainfall may not reach the original source waters.

If, however, water is drawn from subsurface aquifers, it contributes to depletion of those aquifers. Many aquifers are meteorically recharged (i.e., by rainfall percolating through surface exposures of the aquifer formation), but that recharge process is slow. Some aquifers are confined (surrounded by impermeable rocks) and, thus, will deplete just like oil and gas reservoirs (Ransom 2014).

Depleting a potable aquifer for frac water would appear to represent a serious cost that should be avoided. Water will still be released upon combustion, but it will not find its way back into confined aquifers. Of course, whether an aquifer is confined or meteorically recharged, if most of the gas is piped long distances to consumers, then the water released by combustion is lost to the local water system.

However, even with this argument designed to at least partially allay concerns of the water needs of fracturing, it must be made clear that water vapor itself acts as an agent in climate forcing, enhancing the effect from CO_2 in the atmosphere. So even with this apparent balancing of water usage at local and global levels, we must accept that there is a legitimate impact on the overall environment. This is an important aspect of creating robust energy value models: that which is good for one thing (e.g., global fresh water balance) may be bad for something else (e.g., climate change). This will be true for many of the choices we make and analyses on which we base them.

"Fracking" and Ground Water Contamination

Many are worried that the chemicals in the fluids can find their way back to the surface, especially to surface water aquifers. In the Osborn et al. (2011) study, they show that contamination is a problem, but it occurs close to the surface around well bores. This most probably indicates defective casing or cement in the well. If the contamination was related to fracturing, one would not see this correlation since factures are deep underground and reach out a mile or more away from the well.

The problem with focusing on fracturing rather than the integrity of casing is that it ignores what does go wrong and can be fixed. To be clear, hydraulic fracturing is a procedure conducted on some wells after drilling to fracture the reservoir rock to create permeable flow paths for oil or gas to reach the well-bore. The fractures do not extend to the surface. They are confined to within a few hundred feet of the zone of interest, which is thousands of feet below the surface and below any potable aquifers. Indeed, if a fracture were to extend upward from deep reservoirs, at about 3000 feet of depth, it would turn horizontal, due to the dominant stress in the rock.

Even with this clarification, it may be helpful to establish two particular regulations that would likely improve confidence about safety. First, set casing and cement inspection standards to ensure they are of good quality throughout the active life of the well. Research is needed to develop the best practices and the industry is addressing these issues, but this new regulation would inevitably accelerate that process. Second, set a minimum depth above which fracturing is highly restricted. Wells should probably be 4000 feet deep before fracturing is initiated. Regulations should relate the required depth to the size and pressures of a frac job, details that must be informed in consultation with fracture design experts.

Studies have extensively mapped the extent of fracture growth in both The Barnett Shale in Texas and in the Marcellus Shale in the Appalachian Basin. The Marcellus (which is a focal point of much of the controversy about frac jobs) does show higher fracture growth than the Barnett, but do not show fractures growing to shallower depths than 4500 feet. The deepest freshwater aquifers are generally less than 1000 feet deep. This means that the top of the most extreme fracture growths are still separated from potable aquifers by thousands of feet of rock: generally two-thirds of a mile (Fisher 2012). There is no likelihood of any significant amount of frac water traveling up to contaminate surface waters.

Although huge quantities of water are used to comprise the frac fluid, the relative proportions of potentially contaminating chemicals are quite small. The composition of frac fluids is typically 98–99.5% water. The total chemicals used are extremely diluted once mixed into the overall frac fluid. Of these chemicals, the largest fraction is generally hydrochloric acid (HCl). Of course, pure HCl is quite dangerous if humans are exposed to it. By the time it is mixed in frac fluid, though, its concentration is reduced to about one-eighth of 1 percent. The natural concentration of HCl in your stomach is about 4 times greater. Furthermore, the acid that is in the fluid at least in part will react with and dissolve carbonate material in the rock, which helps neutralize the acid (US DOE 2009).

The various other chemicals are mixed at lower concentrations. Most of them can be found in much higher concentrations in household products, which are spilled on the ground by someone somewhere every day. Even if a portion of the frac fluid somehow found a conduit to the surface (such as an induced fracture intersecting a naturally existing fracture or fault line) and enter the ground water, the frac fluid would be further diluted by the ground water. No contamination of drinking water sources with alien chemicals is desirable, but the impact on water quality would be extremely low.

In assessing risk, the most accepted practice is to multiply the severity of an outcome by the likelihood of that outcome. Therefore, since the severity is low (the chemicals are heavily diluted in the original fluid, which is further diluted by the water source it would enter, and some of the prominent chemicals will react with rock materials and be neutralized) and the likelihood has been demonstrated to be exceedingly low, the risk to human safety from the frac jobs is really too low to measure reasonably. The impact of highly diluted and neutralized chemicals times a likelihood that is clearly less than 1 in 10,000 is exceedingly small. The real risk of cleaning your bathtub, or spraying for ants, or painting your house is much higher. The chemicals are in much higher concentrations and the likelihood of an accidental spill contaminating something that you ingest is probably higher.

Of course, it is also possible for chemicals to be spilled during the transport and mixing of frac fluids. This is a much more serious risk factor. If the pure chemicals are spilled before mixing, they may pose real danger, because they are in thousands of times higher concentrations. It is also more conceivable that fluid spilled on the surface would run off into streams or percolate through the soil and surface rocks to contaminate aquifers. It is, then, very important to take multiple layers of precaution to ensure against chemical spills on the surface. Most oil and gas companies are indeed using multiple layers of containment to prevent any fluid spilled from running off or contaminating ground water. The ground is covered with rubber mats or tarps and dikes are built around the entire well site and around each area where chemicals are being handled.

The need for utmost caution in handling chemicals is true for all large-scale activities that use potentially dangerous chemicals (and almost all chemicals are potentially dangerous in high concentrations). Lots of chemicals are used in the manufacture of photovoltaic cells, and they will be used, transported, and stored in vast quantities if solar cell production takes off at the rate that many people hope. It may well be that this chemical intensive process will result in more chemical spillage than frac jobs. It is also true for all large-scale activities that, regardless how many precautions are taken, some accidents will happen and some chemicals will be spilled. We must seek to engineer carefully to mitigate against spillage.

Oil

Another large concern about water use is the process of waterflooding oil reservoirs.[4] It consumes far more water than even fracturing. Still, the argument holds that, when the oil is burned, it will release more water to the atmosphere, but the net gain is smaller. It remains an important (largely unattended) issue to ensure that the water is not being withdrawn from confined systems or from systems that cannot benefit from the recharge process.

In addition to the use of water, it is chemically contaminated by releases from many human activities. Oil spills are among the most notorious, but as we've noted not even close to the largest source of water contamination. Catastrophic spills, though, do have the potential to produce especially severe pollution episodes. Crude oil can impact wildlife, but not through extreme toxicity. Crude oil is not particularly toxic. Many, particularly synthetic products, are quite toxic though. Pollution from millions of tiny domestic, commercial, and industrial processes put seriously toxic chemicals into the water.

Power Plants

Water use by power plants has been linked to water shortages also. This has been noted for many years as a cost of thermal power plants, whether the source is a fossil fuel, a biofuel, or a nuclear fuel. Short-term shutdowns of nuclear power plants in Australia, France, Germany, Romania and Spain have been attributed to drought. California has moved to create limitations on once-through water use, to avoid risks of having to shut down power plants in that water constrained state.

Kent Zammit of the Electric Power Research Institute is quoted as noting that any thermal power plant with once-through water usage will have similar water impacts. The cost is likely to be seen as an externality: Water will not be expensive when it is available, but the risk that a plant's production may be delayed or interfered with could be a major cost (Gies 2010).

---

[4] Waterflooding is a common practice in older oil fields. Water is injected in some of the field's wells to push oil toward other wells. This process commonly doubles the amount of oil recovered without it.

Biofuels

Crops grown to produce biofuels are likely to use water at two stages of the system: irrigating the fields and processing the harvest to produce fuel. We suggest that this is a somewhat neglected cost of biofuel systems. The River Network's Bevan Griffiths-Sattenspiel seems of like mind, suggesting that biofuels are likely to consume anywhere from 2 to 200 times more water than gasoline (Gies 2010).

When the crops are grown in regions that need irrigation, a great deal of water is consumed. True, the water remains a part of the surface hydrologic cycle, but it is commonly drawn from sub-surface aquifers, which means that a water resource is being depleted at a rate that is likely to be greater than its rate of replacement. (Surface water does percolate down through overlying rocks to recharge aquifers, but that is not an instantaneous process; it may take many years.) Just as with the renewal rate of fuels, we must be very conscious of the renewal rate of water within its natural system—especially confined aquifers.

The processing of biofuel feedstocks also uses a great deal of water. Joao Chidamoio, a Mozambican professor, when doing graduate research at the University of Rochester, chose to write his thesis on the feasibility of biofuels for development in Mozambique—an issue with which he had professional experience. His conclusions included that water resources could prove a serious constraint on biofuel development (Chidamoio 2008).

Renewables

If the costs of water demands are sufficiently accounted for, they will likely shift favor toward wind and solar photovoltaic energy systems. Solar thermal projects will have similar water demands to combustion and nuclear power plants. Of course, every manufacturing process needs water, but photovoltaic and wind systems will probably have little or no more water demand in manufacturing than conventional power plants.

These popular renewables are not likely to have any harmful impacts on water systems during their operation (Gies 2010). Geothermal power can be operated on a closed loop cycle, in which essentially all of the steam produced is condensed and re-injected into the reservoir, which would also allow it to operate with minimal water impacts.

Subsurface Impacts

We often do not think of the subsurface as part of the natural environment, but it does play important roles. Much of the water we use comes from subsurface aquifers, as discussed above. The Earth's crust supports all that is on the surface. When we impact it, we impact our environment.

Resources that are extracted from the Earth's crust have a potential to create a variety of impacts on the earth itself. These can include seismic events, subsidence, induced fractures, erosion, and contamination of aquifers. Oil and gas operations have drawn a great deal of recent attention, but some of these impacts are also strongly related to mining for coal and other resources and can be related to large-scale nuclear power plants and hydropower systems.

Society long adopted an "out of sight, out of mind" approach to wastes. Even pollution emitted to the atmosphere or dumped directly into rivers and the ocean seemed to disappear, being absorbed by systems that seemed limitless—or at least too big for humans to impact. Certainly, we didn't tend to worry about anything underground. "Buried and gone." But, of course, that's not true. Even that which we bury remains and can contaminate important systems, especially fresh water (potable) aquifers.

Perhaps the impacts that are the most invisible to us and that we certainly worry the least about are the billions of small releases of waste products by individuals every day. This includes the cleaners, pesticides, herbicides, antiseptics, paints, and even detergents we use. Then there are all of the products that break down or wear out that need to be trashed. All of these products include chemicals, which can be harmful if concentrated. These pollutants are not truly a part of energy systems, but may be incorrectly attributed to energy systems. (Much of what is popularly claimed to be contamination from frac jobs seems more likely to be a result of domestic, agricultural or other waste streams, for instance.) All of these substances can either enter surface or subsurface waters.

Final Thoughts

Since, ultimately, all that we have and are comes from our environment, it must be valued very highly. The challenge lies in the process of assigning comparable values to such a wide range of impacts. Some of the impacts we've seen are dependent on context. We'll see that evaluating environmental impacts will require something of a decision tree approach in order to contextualize the issues properly. It will be difficult to compare environmental impacts, as their impacts on humans are indirect.

We would suggest that, like economics, environmental values, be risk and time discounted. Risk discounting just makes sense. Pollutants are known to be released by all combustion processes, but we also know that more pollutants are released by solid fuels (firewood is the worst, then coal and charcoal). We also know that coal mining has substantial human safety hazards.

Thus, we leave with a few important highlights that exemplify what to be looking for as we set off on new decisions relating to energy systems in this century:

- All activities have consequences.
- The scale of activity is crucial in assessing its impact.
- Pollutants that find their ways into air or water systems are likely to impact animal life, including humans.
- Global climate change from anthropogenic impacts has potential for catastrophic consequences and must not be ignored.
- Other environmental impacts must not be obscured by an exclusive focus on climate.
- The impacts of raw biomass dependence are very large, but too often ignored.
- Land use should be considered in reference to its disruption to ecosystems and to whether it allows for multiple use or competes with other needs.
- Fresh water is a hidden resource cost to energy use. Impacts on water systems should be evaluated thoroughly.

References

Chidamoio, J. 2008. "Life cycle assessment of biofuels in Mozambique." MS Thesis, University of Rochester.

County Engineers Association of Ohio (CEAO). 2012. "Model roadway use and maintenance agreement for horizontal drilling projects and infrastructure." Accessed from: http://www.ceao.org/aws/CEAO/pt/sp/home_page

Fisher, K. 2012. "Hydraulic fracture height growth: Real data." SPE Production and Operations Journal, 27, 8–12.

Gies, E. 2010. "Water adds new constraints to power." The New York Times. Accessed from: http://www.nytimes.com/2010/05/18/business/energy-environment/18iht-rencal.html

Helfman, G. S. 2007. Fish Conservation: A Guide to Understanding and Restoring Global Aquatic Biodiversity and Fishery Resources. Washington, D.C.: Island Press.

Ingraffea, A. 2013. "Gangplank to a warm future." The New York Times. Accessed from: http://www.nytimes.com/2013/07/29/opinion/gangplank-to-a-warm-future.html

International Association of Oil and Gas Producers (OGP). 2000. "Flaring & venting in the oil & gas exploration & production industry." Report No. 2.79/288. Accessed from: http://www.ogp.org.uk/pubs/288.pdf

International Energy Agency (IEA). 2009. "Bioenergy—A sustainable and reliable energy source." IEA Bioenergy.

Karam, A. 2001. "Nuclear energy" in Encyclopedia of Life Support Systems. Developed under the Auspices of the UNESCO. Paris, France: EOLSS Publishers. Accessed from: http://www.eolss.net

Luoma, J. R. 2010. "World's Pall of black carbon can be eased with new stoves." Yale Environment, 360. Accessed from: http://e360.yale.edu/content/feature.msp?id=2250

Mulloy, K. B. 2013. "Occupational Safety and Health." National Academy of Engineering "Shale Gas Promises and Challenges." followed by personal telephone conversation, July 2013.

Nation Renewable Energy and Energy Efficiency Lab (NREL). 2014. "PV FAQs: What is the payback for PV?" Solar Energy Technologies

Program. Accessed from: http://www.nrel.gov/docs/fy04osti/35489.pdf

Niman, M. 2003. "Kodak's toxic moments." Accessed from: www.alternet.org/story/16030/kodak%27s_toxic_moments

Osborn, S. et al. 2011. "Methane contamination of drinking water accompanying gas-well drilling and hydraulic fracturing." Proceedings of the National Academy of Sciences, 8(20), 8172–8176. doi: 10.1073/pnas.1100682108

Ransom, R. 2014. "Interpretive groundwater glossary." Accessed from: http://www.interpretivegroundwaterglossary.com/

Rassenfoss, S. 2013. "Searching for opportunities in environmental trouble." Journal of Petroleum Technology, 65(12), 62–68.

Reist, P. C. 1993. Introduction to Aerosol Science. London: Macmillan Publishing Company.

Smith, K. R. 2006. "Health impacts of household fuelwood use in developing countries." Unasylva-FAO, 57(2), 41.

Stern, J. 2014. "Russian Influence on Gas Supplies." Interview on C-SAPN news network, May 27, 2014.

United States Department of Energy (US DOE). 2009. Modern Shale Gas Development in the United States: A Primer. US Department of Energy, National Energy Technology Laboratory. Accessed from: http://energy.gov/sites/prod/files/2013/03/f0/ShaleGasPrimer_Online 4-2009.pdf

United States Environmental Protection Agency (EPA). 2013. "Mid-Atlantic Mountaintop Mining." http://www.epa.gov/region03/mtntop/

United States Environmental Protection Agency (EPA). 2014. "U.S. Greenhouse Gas Inventory Report: 1990-2013." Accessed from: http://epa.gov/climatechange/ghgemissions/usinventoryreport.html

United States Environmental Protection Agency (EPA). 2015. "National Emissions Inventory." Accessed from: http://www.epa.gov/ttnchie1/trends/

US Energy Information Administration (EIA). 1998. "Natural gas: Issues and trends." Accessed from: http://www.eia.gov/pub/oil_gas/natural_gas/analysis_publications/natural_gas_1998_issues_trends/pdf/chapter2.pdf

US Energy Information Administration (EIA). 2014. "North Dakota aims to reduce natural gas flaring." Accessed from: http://www.eia.gov/todayinenergy/detail.cfm?id=18451

US Energy Information Administration (EIA). 2015. "Annual Coal Report 2013." Accessed from: http://www.eia.gov/coal/annual/pdf/acr.pdf

US Nuclear Regulatory Commission (NRC). 2015. "Backgrounder on radioactive waste." Accessed from: http://www.nrc.gov/reading-rm/doc-collections/fact-sheets/radwaste.html

Upton, J. 2014. Wind Turbine Syndrome Cast Off By Courts As Developers Win Suits Filed By Disgruntled Neighbors. Huffington Post

Accessed from: http://www.huffingtonpost.com/2014/08/21/wind-turbine-syndrome_n_5698211.html

Western Pennsylvania Coalition for Abandoned Mine Reclamation (WPCAMR). 2013. Accessed from: http://www.wpcamr.org/organization/history.html

World Health Organization (WHO). 2014. Household Air Pollution and Health. Report No. 292. Accessed from: http://www.who.int/mediacentre/factsheets/fs292/en/

Zhou, L. et al. 2012. Impacts of wind farms on land surface temperature. Nature Climate Change, 2(7), 539–543.

CHAPTER 6

Social Impacts

Energy Systems in Society

Most of the value generated by energy is measured by how people are positively impacted by using it. Of course, the benefits are closely linked to the sorts of needs the energy serves. Unlike food, though, energy consumption, by itself, is worthless to us. That is, energy systems must be made to conform to the needs they serve.

Even to the extent that energy systems meet needs, there are also social costs with each form of energy. These are often related to the ways in which energy access generates wealth, which is unlikely to be uniformly distributed among all people. Safety hazards associated with energy production and consumption also fall into the category of social impacts.

Consider the following summary list of factors related to the social impacts of energy systems:

- Energy provides for critical human needs and the basis for all development, offering valuable positive social impacts.
- The value produced by energy consumption is linked to the human needs it serves.
- The importance of energy in people's lives stimulates a level of anxiety over its availability.
- Large-scale business activities create job opportunities, which stimulate local economies.
- Extensive operations create demands on local services, which can have both positive and negative social impacts.
- All human activities have risks to health and safety.
- The more energy any system produces, the greater will be its hazards, but some systems carry more intrinsic hazards than others.

All that we experience in modern society relies on energy. We are consumers of all of the products energy is used to make. We are also recipients of the waste products and hazards. Many of the impacts are negative. However, we use energy because of the important benefits it provides us.

For residents of affluent nations, for example, transportation is a prominent benefit sought from energy use. Transportation not only moves us, but brings us goods from all over the world. We would not have the variety of food that we take for granted all year round. We would not have the raw materials, or the cheaply manufactured and produced goods. Indeed, we likewise would not be able to live in some climes without the ability to heat our living and working spaces in the winters. We would certainly not be able to enjoy our entertainment, our communication, even much of the means by which we get information.

Proper valuation of energy systems demands proper attribution of value added, as much as appropriate discounting for costs incurred. The fossil fuels provide a vast majority of the world's energy, which means that they provide a vast majority of the support for many activities essential to survival and for all of our economic productivity.

In the case of petroleum, its value goes far beyond the energy it provides. It is also the feedstock for nearly all synthetic products: lubricants, plastics, and even pharmaceuticals. Coal has the potential of serving as a feedstock for synthetic chemicals as well, even though it is not called upon to do so extensively while petroleum is produced in abundance. But, when oil and gas peak and begin to decline, the value of their non-energy by-products will need to be replaced.

If we attempt to deny the value of oil and gas, and even coal, we deny the basis of modern society. Some may dream of a pastoral existence in mild climates, with little energy input, but there is no way for some 7 billion humans to live on Earth without the support of value-adding energy systems.

Instead of taking them for granted, we should try to understand and respect energy's benefits: to attribute appropriate values to them; yet recognize their limits and their flaws; look to improve; and to move toward more sustainable and even higher aggregate value energy systems.

Social Impacts Related to the Sources

As with all impacts, every source and every system has both positive and negative social impacts. It is important to be clear that we consume energy to provide benefits to ourselves and society. Therefore, it is important that our energy sources add as much value, with as little cost as possible. Essentially, all of the value added is in terms of positive social impacts.

The values added are a function of the quality, versatility, and dispatchability of the energy we use, which varies to some extent by the resource being tapped. Many of the negative impacts that reduce the net value gained by our energy use are felt indirectly through impacts on other systems: resource depletion; environmental impacts; and economic costs. It is now time to consider direct social impacts.

Health and Safety of the Sources

Some of the largest costs to energy systems relate to human health and safety. Energy providers must regard the concerns of people seriously, regardless of whether those concerns are based on misunderstandings or errors. The concerns are still genuine. The issues impacted are still important: clean air, clean water, safety, and livable communities.

While the impacts on air and water and human safety each have demonstrable costs, the social impact also relates to perception. Although perceptions may be erroneous, they are real in their own right. The security of these rights is so critical that it behooves us to take seriously anxieties that are seen in public perception. Living with fear, anxiety, and distress certainly impairs the real quality of life we enjoy.

The Combustion Fuels

All of the combustion fuel industries have rather high rates of fatal and disabling injuries. The rates are most commonly reported per worker hour, not per unit of production. Naturally, these industries have more total injuries, because they produce far more energy and employ more workers than all of the others. We suggest that impacts, costs, and benefits all should be compared on a BTU-delivered basis. That would place the analyses on common ground, making for more meaningful comparisons.

Oil and Gas. The incidence of injury in oil and gas operations is relatively high per worker hour, but oil and gas operations are not very labor intensive, so the incidence is much lower relative to the amount of energy produced. With 25 U.S. fatalities in oil and gas in 2012 (BLS 2013), the incidence is dwarfed by many other human activities.

This is not meant to trivialize the significance of the deadly and disabling injuries in the oil patch. There are significant hazards in the large-scale operations of oil and gas development. The drilling rigs are typically very tall (greater than 100 feet to accommodate "trebles" of 30 foot joints of pipes in the derrick), placing workers at risk of serious falls.

They employ massive forces to drill tens of thousands of feet into solid rock. At these great depths, tremendous pressures can be encountered: even 10–15,000 pounds per square inch. That pressure would be the equivalent of having one to two large vehicles parked on every square inch of your body! Oil and gas are also extremely flammable—that's the very reason we seek them as fuels. But this makes them explosion hazards.

They remain hazardous in processing and to consumers. Contributing to the hazards of oil and gas operations is the fact that wells are constantly being drilled in new places, many of which are remote or in challenging environments. This makes it more difficult to establish standard procedures and workplace practices, as the conditions are so variable.

The Macondo (i.e., Deepwater Horizon) disaster of 2010 demonstrated shockingly the hazards of blowouts with gas, in which the explosion not only destroyed the drillship and led to an enormous environmental disaster, but also killed 11 people. These catastrophes rightly draw considerable media attention. They are rare, for systems that produce staggering quantities of energy.

In addition to the dramatic catastrophes that such massive scale operations can incur, there are longer term, and more insidious hazards associated with oil and gas operations as well as with other energy systems. Although a great deal of attention is focused on potential risks of the (mostly innocuous) chemicals used in modern, large-scale hydraulic fracturing jobs, there is little evidence to support these concerns. On the other hand, hardly anyone discusses the hazards of handling huge quantities of sand as proppant for the frac jobs.

Ultrafine silica particles being released and mobilized in handling the quantities of sand do offer a serious health challenge to workers. Ultrafine particles are the basis of black lung disease notoriously experienced by underground coal miners. The particles are too small to be filtered out or trapped by nasal membranes or natural defense mechanisms in the trachea or lungs. The very tiny particles can enter cells in the lungs, even doing cellular-level damage.

In terms of frac sand, this is a threat that can be addressed rather effectively, with engineering controls and personal protection equipment to limit workers' exposures to the ultrafine particles. Some sorts of shields or enclosures around the conveyor belts and handling equipment could contain a great deal of the fine particles and personal respiratory equipment can limit the amount workers inhale on site.

With all of these problems noted, still the total number of fatalities in oil and gas in 2011 of 25 people pales to statistical (although certainly not personal) insignificance compared with the fatalities associated with our consumption practices. All the while, oil and gas are providing some two-thirds of total energy.

Coal. Extracting massive quantities of coal from underground mines has been an extremely hazardous occupation since it began. In the early days, there were virtually no safety measures available. The earliest coal mines were of the "drift" type: digging back into a hillside into a seam of coal exposed by erosion. The roofs of the mines often collapsed, burying miners alive. Some of the first safety protocols involved propping the roofs with timbers or even rubble—and tying a rope around the waist of the miner so that his friends might be able to pull him to safety in the event of a collapse!

Another hazard quickly emerged as miners needed light: explosion. Although it was commonly referred to as "coal dust" exploding, much of the hazard was actually associated with the methane that is contained in coal beds. Methane is highly explosive in air with the smallest ignition source. The open flame lanterns that prevailed until the 20th century provided more than adequate ignition. Lanterns were designed to be sealed, with air pumped into the lantern through water, so that no large explosion would occur, but these measures were only as good as the care that miners took not to open the lantern in the mine for repairs. Modern battery-powered lanterns have helped to reduce the risk of explosion and fire.

As mines reached deeper into the earth, air circulation became an issue, resulting in flooding. The mines were often in contact with aquifers and could easily flood. Indeed, the first steam engine applications were to run pumps to remove water from mines. Even into the 20th century, mine flooding has been the source of disasters, but improved pumping and mine design to provide evacuation sites have helped to mitigate these problems.

One of the hardest problems to address is commonly referred to as "black lung disease." The fine particles inhaled working in the mines can be deadly. The particles are very difficult to filter and can actually do damage at the cellular level in the lungs. Innovative respirators can help to mitigate this, especially when combined with very good ventilation. The incidence has diminished dramatically, but the problem cannot be realistically eliminated altogether.

Surface or strip mining ameliorates the risk of flooding and explosion and reduces the risk of respiratory disease by being in open air. Hazards remain though. People are still working in the vicinity of exceptionally massive equipment. Of course, in spite of air circulation, the mining process liberates vast quantities of fine particles.

Without accounting for long-term respiratory health effects, U.S. coal mining experienced 20 fatalities per unit of energy produced in the years 2011–2013 of 20 (MHSA 2015). That is slightly less than the 25 for the petroleum industry, but remember that oil and gas together produce twice as much energy as coal.

Firewood. The hazards of coal mining, though, are utterly dwarfed by the hazards of raw biomass dependence. The World Health Organization has been steadily increasing their estimates of fatalities from firewood use over the past decade. Even as we have been writing this book our use of a relatively recent estimate of 2 million firewood-related deaths per year has now been increased to over 4 million deaths per year.

All of the other energy sources add up to a small fraction of the death toll from this one energy system. These firewood-related deaths are primarily attributed to smoke inhalation. There are also fatalities associated with gathering the wood (from machete cuts, snake bite, and rape and murder of women going into the forests to collect wood). This is clearly the worst energy system in terms of social value—by a large margin.

Nuclear Fuel

Anxieties about nuclear systems are particularly evident and are serious constraints to the growth of nuclear fission. The fears relate both to radiation hazards and to the relationship to weapons. The fear of radiation hazards makes many people reluctant to have nuclear power plants located near their homes and communities. The fear of nuclear weapons makes people—and nations—oppose nuclear power development in many developing nations or those who might be viewed as potentially hostile.

The fear of radiation is clearly exaggerated. Radiation prompts fears, perhaps in part due to its invisibility. Some of the fears are based on erroneous notions. Radiation is physically different from chemical toxicity, but the effects are comparable. Certain doses may be harmless, while larger doses may cause illness and even death. Most chemicals are invisible to us as well.

People often refer to the half-lives of some radioisotopes with alarm. The half-life does give some indication of how long a radioactive substance can be hazardous, but this should not be viewed as making radiation worse than chemical toxins. The half-life shows how the danger diminishes with time. After three half-lives, only one-eighth of the original radioactivity is left. On the other hand, most chemical toxins will be as hazardous a million years from now as they are today.

Radiation hazard is probabilistic, but so are chemical and infectious hazards. Different people have different sensitivities. Where and how the doses are received also affect the degree of damage done. We would suggest that people think about radiation hazards in very much the same way that we think of chemical and infectious hazards. Perhaps the greatest anxieties about nuclear hazards stem from the clear—and monstrous—destructive power that has been demonstrated by nuclear weapons. Is nuclear power not like nuclear weapons?

Processing fissile material does have some similarity to processing materials for nuclear weapons. They both involve acquiring and refining nuclear fissile material, such as uranium. The naturally fissile isotope of uranium is less than 1 percent of the total amount of uranium. Purifying fissile uranium is particularly challenging, because both the fissile and nonfissile isotopes of uranium have identical chemical properties, since they are the same element. The only difference is the nuclear mass.

Therefore, the separation process can be based only on the tiny density difference between fissile uranium. With sophisticated separation technology, the two can be separated fairly effectively—sufficiently to make nuclear fuel. It requires even much more sophisticated technology, though, to enrich (separate) the isotopes well enough to make weapons grade uranium.

Controversies have raged numerous times over whether some nation's nuclear program is actually a nuclear power project or a weapons program.[1] It takes only a modest amount of enriched uranium to make a bomb. The bomb dropped on Hiroshima contained 140 pounds of enriched uranium, but was very inefficient. The subsequent bomb dropped on Nagasaki, was much more effective, requiring less than 14 pounds of highly enriched plutonium (about the size of a softball) although it used thousands of pounds of conventional explosives to drive two segments of the nuclear fuel together (Atomic Heritage 2013). Today, there is concern over the possibility of "backpack" nuclear bombs.

Although it takes massive investment in technology (and of course, access to fissile material), making a nuclear weapon is quite conceivable for many nations. The other big technological hurdle for nuclear warfare is the delivery system. (Systems are important even in weaponry.) With very small nuclear weapons, it is possible for an individual to carry it to the target. There is probably reason to worry. Certainly the specter of nuclear weapons makes people very anxious. We believe that these fears will limit the deployment of nuclear power throughout most of the world for many years to come.

The fears of accidental release of radiation also inhibit nuclear power expansion in those already nuclear nations. Nuclear power growth has completely stagnated in the United States for 40 years. The near disaster

---

[1] Since 2013, Iran's nuclear program has been at the center of such a controversy, but they are not alone. The question has arisen before and will doubtless arise again.

at Three Mile Island triggered resistance, even though no serious injuries have been attributed to it. Before time could mitigate these fears, a real disaster emerged at Chernobyl in the Soviet Union in 1986. About 30 people died promptly in that accident.

There was tremendous anxiety caused by the radiation cloud spreading across northeastern Europe. Ongoing studies have revealed a few hundred confirmed cases of radiation-related sickness (WNO 2013a). By 2011, even the United States seemed to be ready to take the plunge with nuclear power again. Then a massive earthquake in Japan crippled a series of nuclear reactors at Fukushima. Three workers were killed during the earthquake. No fatalities have been confirmed from the significant radiation released (WNO 2013b).

In spite of the evidence to the contrary, the anxieties remain strong and generate a great deal of speculation. The fears are not entirely unfounded, as a great deal of radiation was released in both of these nuclear disasters. Radiation certainly presents a health hazard. Yet its hazards are rather mysterious, as the radiation cannot be perceived by human senses (unless it is so intense as to cause tissue burns). Fear of the unknown is inherent in human nature. The plagues must have been terrifying to earlier people, not only because of the widespread death, but because the cause was unseen, unfelt, and unknown.

Now, like disease, science has presented an understanding of the hazards of radiation, but the risk and the damage remain probabilistic. The kinds of radiation levels that spread beyond the immediate vicinity of a nuclear accident do not produce prompt, direct death, but rather cellular damage that can lead to long-term health effects, such as cancers. These disorders may take many years to develop and are exceedingly hard to trace to a primary cause. Was it radiation? Was it a lifetime of smoking? Was it diet? Was it a host of other carcinogenic factors?

So, unlike plague and other infectious diseases, there remains uncertainty: mystery. It may be that more people are dying from the radiation exposure than can be determined at this point. Or it may not. The uncertainty is a social cost of nuclear power.

Of course, nuclear power plants of the late 20th to early 21st century have been very large scale, which carries with it the potential for large-scale impacts concentrated around the sites of the nuclear reactors. Everyone wants the electricity, but almost no one wants a nuclear power plant in their "backyard." It doesn't suffice to dismiss these (or any other) anxieties. There must be real, honest communication between an anxious public and government officials, and professionals with expertise.

Waste disposal remains another important concern and one that continues to be too controversial to reach a resolution in the United States. It seems rather curious, in light of the fact that radiation is dangerous if it is concentrated, that the plans for handling high-level radioactive waste is to concentrate huge quantities of it in a selected site. Of course, that site is selected to be safe—to provide a secure repository for the waste for many years. However, it may be well to remember that much remains beyond our ken. No human planning is infallible. To the extent that massive quantities of highly radioactive material are to be concentrated in one place, even if the probability of failure is exceedingly small, the consequences of failure could be monstrous.

Nuclear fission will not be likely to meet a large share of world demand because of both the rational and the irrational fears that create real social costs. These fears need to be addressed. Most people are not likely to have a rare form of cancer. However, if we approach our doctor with a concern, the doctor needs to ask us questions about our concern and find a methodical, scientific means to rule out that disease—and perhaps find some real disorder to treat.

Nuclear power for civilian use has been done at a very large scale. That scale contributes to the anxieties, as it invokes the potential for catastrophic failure that may impact people far beyond the workers themselves. The data strongly suggest that the total number of human deaths is far less in nuclear power than in most energy production activities, but that must be taken in context of the reality that there have been some large-scale failures, with large releases of radiation. People have reason to want serious, thoughtful reassurance that safety concerns are taken very seriously. It is nonsensical to dismiss the reality of hazards associated with large-scale energy activities.

"Renewables"

Once again, it seems worth emphasizing the point that the rate function that determines the relative "renewability" of different sources has nothing to do with the health or safety impacts. We did, therefore, categorize the "combustion fuels" together in the previous section. In this section, we will discuss the noncombustion resources that are popularly referred to as renewable.

Hydropower. Hydropower is, by far, the largest producer of the potentially renewable energy sources. Much of the hydropower in the United States was developed as part of the Tennessee Valley Authority (TVA), in which electricity generation is only one of the target benefits of the program. The TVA website says "TVA built dams to harness the region's rivers. The dams controlled floods, improved navigation and generated electricity." (TVA 2012). It was a true systems-level approach, seeking address multiple resource and human issues.

Floods devastated farming in the region, where malaria threatened lives. Dams were very effective at these purposes and the TVA continues today. Hydropower was extensively developed across the United States (as well as much of the Developed World), often touting recreational benefits as a positive social impact.

Nevertheless, numerous negative impacts have inevitably emerged as well. Big dam systems dominate the generation of hydropower and they have a clear potential for accidents with high death tolls when they fail. This is a relatively rare incidence, though. When dams do fail it can be catastrophic. Indeed, the fatality data are heavily influenced by a single catastrophic dam failure in China in 1975, killing 26,000 people (NEEDS 2008).

The most prominent of the social impacts is the potential displacement of many people from lands about to be flooded by the reservoirs behind the huge dams. The Three Gorges Dam in China drew immense controversy for the displacement of about 1.2 million people (International Rivers 2013). However, the electricity generated by the world's largest dam clearly provides important benefits to Chinese society, opening the door for impressive economic growth and development. That development demanded tremendous quantities of new energy in China, and the dam was one way to provide a great deal of that energy. If they didn't build the dam, how would they have put that much energy into the system to support development so quickly?

We argue, consistently, that it does not suffice to say Developing Nations should not be allowed to promote development. The Brundtland Commission's report that launched the popular discussion of sustainability called for meeting the needs of today's people, without sacrificing opportunities for future people. Perhaps we must, more realistically, seek to minimize the future sacrifice, while meeting the needs of current generations, hoping that advances in technology can mitigate that sacrifice.

Hydropower will no doubt continue throughout the foreseeable future. It adds value. Perhaps smaller scale, "run-of-the-river" projects, which don't impound water behind a large dam will take a larger share, as social (and environmental impacts become increasingly valued.)

Wind and Solar. Wind power was touted in the late 20th century as a very high-value, renewable energy source. The system produces no air or water pollution (if you ignore the manufacture and installation of the turbines.) Soon after large wind power installations began to appear, though, the popular impressions began to change. Some direct environmental impacts were noted, but the large turbines looming on the horizon have drawn considerable animosity.

De gustibus non disputandum est: there's no accounting for taste. People will travel thousands of miles to admire the old Dutch or Spanish windmills, but some of the same people are disturbed to see modern wind turbines. As wind production has grown, its potential negative aspects have been seen and it has lost favor in the eyes of many people. Perhaps the same will be true of the other favored alternatives.

Solar power seems to be the most beloved of the alternatives at the current time. It will probably remain a preferred alternative well into the future, although it may lose some favor as it moves into large-scale production. We suggest that it currently benefits from some "rooting for the underdog." Once it does establish itself as a major energy system, creating great wealth in the hands of the few who bet on the ideal systems, will people start to speak derisively of "big solar"? Perhaps not, but there will be impacts that people begin to recognize and the dream will turn into a reality that is not as utopian as people hope. (OK, we are making a bold prediction here, but it is in keeping with many other systems that have grown to dominate global markets.)

Hazards will be associated with scaling up wind and solar activities. Considering that wind and solar power produce such a tiny fraction of energy, even a single fatal accident per year equates to more fatalities per unit of energy produced than the fossil fuel industries. Although the solar industry is not the subject of separate statistics, it can be seen that the state of California alone investigated more deaths than this in the solar industry: three in two years (Fair Warning 2010).

Accidents happen in every human activity. The scale of the activity is crucial in determining the scale of the threat to human lives and safety. It is fair to say that dramatic increases in solar and wind power production will see their own increases in fatalities in those industries. None of this should suggest that the expansion of these alternatives should stop. We merely need to understand the costs that attend all of our activities. There is no free ride.

Wind has been growing faster than any other noncombustion energy source. (Solar may have 1- or 2-year percentage growth rates that exceed wind's, but since wind has a much larger existing production base, wind's absolute growth has been significantly larger.) As wind scales up, we see the massive towers, hundreds of feet tall. They require massive equipment to setup, applying forces. The height and the forces required for installation, for maintenance, and for ultimate decommissioning place workers in harm's way. We can be confident that fatalities will occur, perhaps less than some of the current mainstay sources, but probably in a similar range for the amount of energy produced.

No Activity Is Without Risk

Every human activity incurs risks. Indeed, Shakespeare gave Julius Caesar a line about death, as a necessary end, coming when it will come. We all die. When and how we die is the question. There is no activity without risk, though some have great risks than others. Bear in mind that driving a car is one of the most ubiquitous American experiences and one of the greatest killers of Americans.

Every energy system requires manufacturing, travel for workers, manual labor, and handling chemicals and materials. The larger the scale of these activities, the more exposure people have and the more accidents there will be. Make no mistake. Part of the reason that there are many incidents in coal and oil and gas operations is the massive scale at which they operate. As new energy systems grow to global scales, they too will have their share of accidents.

Consumption Hazards

Another elephant in the living room is not related to producing energy, but to how we consume it. Some 40,000 Americans die annually in traffic accidents. Globally the number is over 1 million per year. That clearly places our consumption pattern in a strong second place in terms of direct hazards to human lives. This hazard is only to a source insofar as petroleum has proved capable of producing enormous amounts of energy at very low prices.

As long as we consume vast flows of energy to support hundreds of millions of individual passenger vehicles scurrying around our highways, there will be collisions and there will be fatalities. Whether the cars are propelled by gasoline internal combustion engines or by batteries or even by fuel cells, fatalities will remain about the same. It is the number of vehicles, operated by so many people that leads to the fatalities.

However, there are policy and technology changes that can reduce the incidence of fatal accidents. A number of design factors and policies can mitigate the risks. Seat belts, followed by air bags, have significantly reduced the severity of accidents, especially in affluent nations. There is some indication that innovations in automation may be close to introducing automobiles that are operated by computers and electronic sensors: driverless cars. This offers a possibility of eliminating accidents caused by driver error, inattention, and distraction.

A larger potential reduction in fatalities could be provided by large-scale shifts toward modern public transportation systems. There will still be accidents (all systems, of every kind experience accidents). There will be loss of life in some accidents. Nevertheless, public transportation puts fewer vehicles on the roadways, with more controls. Some run on rails.

In addressing traffic fatalities, we must address a controversy about causes of traffic deaths. Many people believe that larger vehicles protect their drivers, even arguing that small (and thus more efficient) vehicles kill. This line of thinking appeals to an intuitive sense that size affords protection, but flies in the face of real data—and a more serious look at the physics of collisions.

Do bigger vehicles offer greater protection? One argument for larger vehicles goes so far as to claim that the push for fuel efficiency and smaller cars killed "7700 people for every mile per gallon gained" (USA 2001). This is indeed a grievous charge, if it is true. So, did fatalities jump dramatically after the United States embarked on a transition toward smaller more efficient vehicles in the 1970s? In fact, actual data on traffic fatalities seem to defy this notion. The National Highway Traffic Safety Administration (NHTSA 2012) shows that 2010 had a record low incidence of fatalities per vehicle mile driven.

Indeed, traffic fatalities peaked at about the time that new fuel efficiency standards were beginning to be implemented and have generally been declining since. Is that possible, if it were true that those standards led to an additional 7700 deaths for every mile per gallon improvement? If that claim were true, would we not expect fatalities to be even much higher in countries that have a greater abundance of small, efficient vehicles? Yet, that is not true either. More efficient vehicles are no less intrinsically safe than less efficient ones.

Good analysis should be based on deaths per mile driven. Furthermore, if larger vehicles are safer, one would expect fewer fatalities per collision of two large vehicles than for two small vehicles. The numbers do not support this either. In fact, the probability of fatalities per mile driven are (very slightly) higher for the larger vehicles (light trucks) than for cars. The data suggest that concerns are totally unfounded regarding smaller, more efficient vehicles being less safe. Our review of data implies that the real hazard comes from having large mismatches of vehicle sizes on the road. If most vehicles were small, efficient ones, it appears that we would all be safer.

Some might say, "Well that's just nonsense. Simple physics tells us that bigger things win in collisions." There are several misapprehensions in this thinking. First, we do acknowledge that if two objects of similar construction collide, the smaller one is likely (not guaranteed) to be more severely damaged, but that refers to the vehicle itself, not the occupants. Second, it ignores that many other factors are involved in the survivability of collisions: how the vehicle body deforms around the passenger compartment; whether the vehicle tends to roll over; and whether there is any fire or explosion. The physics are neither simple nor direct.

Comparing Fatalities

Table 6.1 is provided to summarize the direct risks/hazards associated with energy systems that we compiled from the World Health Organization's Global Health Observatory documents and from the NEEDS report (NEEDS 2008; WHO 2014). It is clear that the two major causes of fatalities related to energy systems are firewood and traffic accidents.

This should make clear that, while all forms of hazards should be deliberately minimized, the most costly ones (i.e., fuelwood dependence and vehicular traffic accidents) remain strikingly high. Shining light on the firewood issue in particular is very important, because it really does inhibit global progress and equity. Moreover, the greater hazards are driectly attributable to how we consume energy. We can, with strong resolve, change these numbers.

***Table 6.1** Summary of average annual death toll due to direct energy systems risks/exposures.*

Activity	Annual Death Toll
Cooking over open wood fires	4,300,000
Driving Accidents	1,240,000
Coal mining (incl. black lung deaths)	20,000
Hydro	857
Oil-related accidents	600
Natural gas accidents (incl. LPG)	200
Nuclear	1

Addressing Present Needs

The scale of activities impacts people and communities beyond overt hazards. In order for energy systems to offer full value, they must be accepted. This is a function of perception, as discussed in the Safety section. It is also a function of how effectively the energy system meets human needs. Of course, needs also involve perception as well.

The primary needs dictate the forms of energy that will be most likely accepted. We harbor reservations about hydrogen fuel cells for cars, in part, because the commonly preferred proton exchange membranes and high-pressure hydrogen storage do not seem well suited to broadly disseminated use in private vehicles. Even electric vehicles have not gained wide acceptance as of yet, in large part because the limited range does not meet the expectations that many people have for transportation. Nevertheless, charging stations are springing up, including at businesses for employees and commercial parking lots (carcharging.com).

In spite of the observation we would offer that good public transportation can be more amenable, safer, and more efficient than private vehicles, the acceptance in the United States is limiting its deployment. Americans certainly do not seem likely to relinquish the perceived freedom afforded by the opportunity to get into one's car and go anywhere at any time. The reality that public transportation does not preclude car ownership does not offset the resistance. Elections have been won in U.S. cities and counties by running in opposition to "smart growth," which includes public transportation.

Part of the opposition relates to taxes and not wanting the government to spend more on the infrastructure required to support modern high speed rail and urban public transportation. Roadway construction and maintenance, however, are large draws on public coffers. If the United States were to divert a modest share of highway funds to effective public transportation, it can start progress toward safer, more efficient transportation systems.

Those of us who have experienced good public transportation tend to be advocates of it. High-speed rail in Japan or Western Europe is fast, efficient, and reliable. Good urban and regional public transportation offers the same benefits. If I am taking a train to work, I can read, or write a book, or take a nap—all with no chance of killing myself or my fellow citizens. It actually frees the rider from the mundane task of driving.[2]

Perception, though, lies at the heart of the challenge. In any context where many of the people view public transportation negatively, it will be difficult to gain acceptance. People need to believe that the system will be readily available, clean, safe, fast, and punctual. Many American public transportation systems have been sufficiently neglected as to defy giving those positive impressions.

In the Developing World

Cooking is the dominant energy demand for half of humanity. Energy systems that fail to meet people's expectations around cooking are not likely to be accepted. This includes time factors and taste. It may be tempting to think that poor people should not be concerned with flavor, but remember that taste evolved, at least in part, as a sense to tell us what is good to consume and what isn't. If food doesn't taste as you'd expect it to, that may be a warning sign that the food is tainted.

Beyond the flavor, food must be ready at meal times. We can take that for granted in the Developed World, where cooking is done at the turn of a dial. But that is not the case for billions of people. In many low-income regions of the world, the rural poor still depend on subsistence farming.[3] The adults often work in the fields all day and the women cook their meals after dark. Thus, solar cookers would be worthless to them. The energy to cook must match the time when people cook. It may be an intractable issue in which the adults are absolutely needed in the fields, or it may be a matter of custom. Customs can change over time, but not overnight.

Even if the people had found a viable way to use the solar cookers, they may well have been dissatisfied with the product. Food heated by

---

[2] Cars are being developed that are designed to be able to use sensors and computers to drive themselves. We doubt that these will advance to the point that the driver is truly free to read or nap, rather merely aiding the driver to avoid accidents for the foreseeable future.

[3] Subsistence agriculture does have its benefits. Although the people work hard every day to grow enough food for their own needs and perhaps a little extra to sell, they do have control over the food they need to live.

solar energy will lack any of the smoky flavors imparted by cooking over wood fires. This same factor tends to work against prompt acceptance of electricity for cooking in previously firewood-dependent regions.

Another cultural factor is that cooking is often a major part of a woman's role in the family. If new energy systems allow much faster cooking, ideally freeing up women's time for more productive activities, it can raise doubts about her role. Spending hours each day preparing meals over open fires has come to be seen in some traditions as a sign of a woman's care for her family. Even for natural gas, which can still cook with a flame and can be designed to impart smoky flavors (as with western gas barbecues), the shift in time may not be initially accepted within some communities.

None of this means that transitions are impossible, nor that they should not be tried. The cultural and social factors, though, affect perceptions of the value derived from the activity that energy supports. They cannot be ignored.

In the Developed World

For the already industrially and economically Developed World, energy systems must be able to address the needs they are meant to serve. Those needs (or desires) are considerably more diverse for more affluent societies.

Basic needs are taken granted to a large degree, but remain essential to survival. Indeed, since many affluent nations are in cooler climates, space heating becomes as important as cooking. The ready presence of a variety of kinds of appliances enables affluent consumers to use many different types of energy sources for cooking and space heating.

Electricity is the most versatile for a wide range of needs. It can be used for cooking and heating, as well as for many applications for which no other energy form is truly applicable. Any energy source can be converted into electricity, with some efficiency cost. In spite of its versatility, electricity has limitations in some applications. Although it is perfectly possible to cook, and to heat water and living spaces with electricity, some people prefer combustion sources. Natural gas is both quick and minutely controlled in modern cooking applications.

Thus far, electric vehicles are failing to achieve significant market adoption, due to difficulties in addressing the perceived needs in vehicular transportation, particularly in terms of cruising range. Although most car trips are relatively short, people desire the ability to get in a car and travel as far as they choose, when they choose. The energy and power densities of gasoline and diesel are very unusual in this regard. There are some hopes for hydrogen-based fuel cells, but there are limitations in terms of storage. Compressed natural gas has potential to achieve greater acceptance for transportation. The pursuit of this option follows on the advent of massive gas resource potential in the shale plays.

Choosing to Adopt New Systems. In spite of all the arguments that can be offered on any topic, if the public is not persuaded to action, nothing will happen. Some of the arguments we make in this book (and that others make) may well be moot, if they aren't adopted at behavioral levels. An effective transition will require placing value on resources, sustainability, and equity to motivate our transitions.

Climate change may be the most noteworthy. There is a great deal of good science that shows that anthropogenic emissions of greenhouse gasses, coupled with destruction of carbon sinks, is altering the global climate. There is clear and present evidence that the alterations could have catastrophic consequences for the environment and for society. The issue is much discussed and has motivated a great deal of emotion.

Nevertheless, we doubt that it is motivating significant behavioral changes. Many governments have signed accords and some policies are being implemented, but behaviors have only changed slightly. Many people still seek to buy homes that are remote from their work or school, while purchasing large, powerful, but inefficient vehicles. Electronic devices continue to proliferate. Homes in the United States tend to become larger, even though families are becoming smaller. Why?

We think that cigarette smoking offers a useful analogy. Excellent scientific evidence has abounded for decades, showing that cigarette smoking is directly harmful to one's health. It ages you and greatly increases the risk of heart disease, emphysema, and a variety of dreadful cancers. Still, cigarette sales remain substantial throughout the world. Young people choose to begin smoking every day—no every second. We suggest that no amount of facts, data, or scientific analyses have been able to eliminate this behavior. Neither of us is psychologically trained, so we won't hazard guesses about the cause. We will just observe that there is evidence that no amount of information will fundamentally change the energy consumptive behaviors that lead to climate change.

We believe that only dramatically increased price will accomplish that. When energy becomes expensive, as we believe it will in the peak oil era, then people will treat energy more thoughtfully and use it more frugally. Only high prices seem likely to achieve the fundamental behavioral shifts that will significantly mitigate climate change and open the door for noncombustion alternatives to compete broadly.

Similar points may be made about other energy debates. No amount of evidence that well-bore integrity of existing wells is the real culprit for water contamination will allay anxieties about "fracking." No amount of evidence about the safety of nuclear energy will suffice. No amount of information about climate change will convince skeptics either.

Equity

Since energy supports development and is a major portion of many economies, it plays a role in socioeconomic fairness. Access to energy supports equity, and energy profits generate great wealth, which offers the potential to lift many millions of people out of poverty. Unfortunately, this potential is not often realized. If the wealth generated finds its way into the pockets of a few powerful people, it can widen the gap of inequity.

We suggest that one of the challenges to equity is that vast wealth is always likely to attract greedy and corrupt people. It always has and probably always will. The rule of law can prevent the most flagrant forms of theft, but it cannot prevent wealth from accumulating in the hands of the already wealthy and powerful.

Too often, in the extractive energy industries, energy is produced from one region, whose populace sees staggering wealth and value being drawn from their lands and shipped away, without benefitting from it. In addition, the people there may be subjected to a number of costs associated with the activities.

Nigeria offers an extreme example. In the Port Harcourt region, offshore oil fields have produced vast quantities of oil for the export markets for decades. A string of corrupt leaders siphoned billions of dollars of wealth from a nation that has a great deal of abject poverty. The energy is shipped to consumers in Affluent Nations. The money was taken by the leaders. The local people are left with the pollution from leaking pipelines and the light from giant gas flares lighting their night, while they have very limited access to modern energy.

Such extreme inequity promotes antagonism. That antagonism, ironically, further exacerbates the negative externalities of the oil field development. Disgruntled people can be drawn to acts of violence and sabotage. The sabotage, some claim, is a major factor in the leaking pipelines (while others claim that it is inadequate maintenance of the field facilities). Inevitably, the atmosphere of violence cannot help but contribute to the problems.

In these times, the giant flares are especially troubling. Many billions of cubic feet of natural gas are being burned off as waste product. Of course, this contributes to global warming, while not offering any value from the energy content of the gas, the cleanest burning combustion fuel. The waste is hideous, especially so tightly juxtaposed with the abject energy poverty of the neighboring communities. This is an environmental and resource cost that should be taken up by environmental activists, to which we believe companies and industry professionals can respond supportively. Engineers abhor waste, and we can work together to abolish this wasteful activity.

Corporate Social Responsibility

Desires to see the great investments and wealth generated to provide some benefit to the local residents, prompted efforts that have turned into Corporate Social Responsibility (CSR) investments. The earliest responses involved local hiring requirements. International companies coming into a host country were required to hire certain numbers of their employees from among the local populace. This was a step in the right direction, but companies often found it easiest to hire locals as cooks, housekeepers, and chauffeurs for their expatriate workforce.

Local investment requirements came along. Ideally even relatively small fractions of the tremendous investments being made by international oil companies (such as 1 percent) could transform local economies and support development. Unfortunately, the oil companies brought expertise in oil and gas operations, not in development studies. Many of the local investments have done little to stimulate human development. A classic example is building soccer fields for children: a nice photo opportunity of smiling children with soccer balls, but no real contribution to development.

Indeed, even the agencies charged with international development do not have particularly impressive track records. Part of this is, we suggest, a general failure to comprehend the interconnection of development needs. It has been all too popular to invest in one or two aspects of development, without supporting the other pieces simultaneously, which are necessary to support sustainable development. Energy was often left out of the development equation completely. When Ben launched a nonprofit effort to support energy for development, administrators at the university where he was working at the time sought an opinion on the endeavor from a major development agency. Their response was that energy was not viewed as a development issue at all. That agency, as well as others, since created energy programs. The importance of energy was finally acknowledged.

Unfortunately, decades of development investments have gone into building hospitals and schools with no access to reliable, modern energy. A friend and colleague passed along an anecdote from a friend of his working in Doctors Without Borders. Lights went out in the hospital in the middle of surgery. This hospital had not only a connection to the electric grid, but also a backup generator, because the grid was not reliable and the hospital recognized the need for energy. However, their generator was not in good repair and, as we heard the anecdote, the surgeon had to temporarily close the patient in order to help get the generator working and lights restored in time to save the patient's life. Whether we got any details wrong in the retelling of this story, it serves to make a point. Hospitals need energy not only to light surgeries, but also to sterilize instruments; to refrigerate medicines; to run diagnostic equipment; and the list goes on. (How many modern physicians want to work without computers?)

Similar points can be made for schools. Lights are essential if anything is to be done on dark days or in the evenings. Computers and audiovisual equipment are nice to have. Indeed, every development activity has energy requirements.

Final Thoughts

We laid out the case in earlier sections that a quantifiable relationship between development and access to energy exists. Energy has gained recognition as a crucial element, but what other elements are not being adequately considered? Furthermore, development is not merely a cake batter in which all the ingredients must be present: it matters how they interact with each other. We suggest that much more work is needed in creating and optimizing sophisticated development models in order to optimize development outcomes. (In this case, we don't necessarily mean that the math must be sophisticated, but that the qualitative relationships described by the models should be.)

Other problems with development are more insidious than ignorance. It has been argued that some development investments are cynically motivated: either to create and support international allegiances or to provide markets for products produced in the "donor" nation (Free Africa 2003). If the actual priority is not to promote development, then that is unlikely to be the outcome.

We suggest that a great deal more attention is needed to understand the complex tapestry of needs and resources that contribute to development. Energy needs to be included in the mix of development planning. Also, it needs to be targeted to crucial development needs and integrated with other activities, using the highest value energy systems for a given context.

To that end, energy systems must be dependable. To the extent that they are not dependable, activities that rely on those energy resources are hampered, if not outright impossible. Energy must be available when, where, and how it is needed, since it is essential to all of our activities. When higher value energy is not sufficiently available to meet essential needs, people fall back on lower value systems, which are likely to have higher social and environmental costs.

The raw biomass dependence of half of humanity is an example. The lack of access to modern energy constrains people to polluting, inefficient, and hazardous fuelwood. When deforestation (spurred by fuelwood harvesting) makes fuelwood less available, the next step down the energy ladder is brush and crop residues, followed by dung. This descent to lower grade fuels stymies development, impoverishes soils. Better energy sources are necessary to improve quality of life.

References

Atomic Heritage. 2013. "Fat man and little boy." Accessed from: http://www.atomicheritage.org/index.php/component/content/article/42/146-fat-man-and-little-boy-bombs.html

Bureau of Labor Statistics (BLS). 2013. Accessed from: http://www.bls.gov/news.release/cfoi.t02.htm

Carcharging.com "Electric vehicle charging services." Accessed May 25, 2015 from: http://www.carcharging.com/

Fair Warning. 2010. "Solar installer's death points to job hazards in a growing, green industry." Accessed from: http://www.fairwarning.org/2010/10/solar-installers-death-points-to-job-hazards-in-a-growing-green-industry/#sthash.lJYt2IPs.dpuf

Free Africa. 2003. "The failure of world bank programs in Africa." The Free Africa Foundation. Accessed from: http://freeafrica.org/wp-content/uploads/2014/02/THE-FAILURE-OF-WORLD-BANK Summary.pdf

International Rivers. 2013. "The three Gorges dam." Accessed from: http://www.internationalrivers.org/campaigns/three-gorges-dam

Mine Health and Safety Administration of the US Department of Labor (MHSA). 2015. "Coal Daily Fatality Report - June 9, 2015." Accessed from: http://www.msha.gov/stats/charts/coaldaily.php

National Highway Traffic Safety Administration (NHTSA). 2012. "Traffic safety facts data 2010." Accessed from: http://www-nrd.nhtsa.dot.gov/Pubs/811630.pdf

New Energy Externalities Developments for Sustainability (NEEDS). 2008. "Final report on quantification of risk indicators for sustainability assessment of future electricity supply options." Accessed from: http://www.psi.ch/ta/NeedsEN/RS2bD7.1.pdf

Tennessee Valley Authority (TVA). 2012. "From the new deal to a new century." Accessed from: http://www.tva.com/abouttva/history.htm

USA Today (USA). 2001. "Deaths by the gallon." Accessed from: http://www.vehicle.org/cafe/usa1.html

World Health Organization (WHO). 2014. "Global health observatory." Accessed from: http://www.who.int/gho/en/

World Nuclear Organization (WNO). 2013a. "Chernobyl Accident 1986." Accessed from: http://www.world-nuclear.org/info/Safety-and-Security/Safety-of-Plants/Chernobyl-Accident/

World Nuclear Organization (WNO). 2013b. "Fukushima Accident." Accessed from: http://www.world-nuclear.org/info/Safety-and-Security/Safety-of-Plants/Fukushima-Accident/

CHAPTER 7

Framework for a New Value Metric

A Possible Future

Figure 7.1 illustrates a transitional path of a possible evolving energy mix until the year 2050. It is based on a commitment to eliminate extensive fuelwood dependence and reduce coal, the worst of the fossil fuels. Relatively optimistic renewable energy growth rates were assumed: 10 percent per year for wind and 15 percent for solar. Similarly, a 10 percent growth in geothermal production begins in 2020. Hydropower production is also assumed to double by 2050. If oil, natural gas, and nuclear are held constant though, a deficit (shaded in black at the top of the graph) rather quickly develops between total energy production and growing demand through 2050.

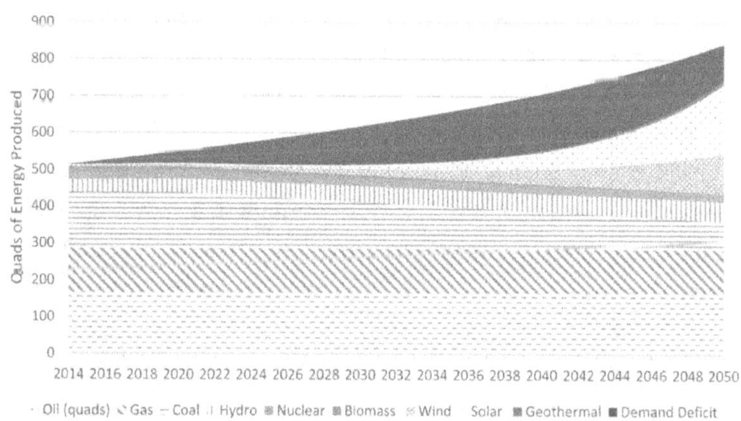

***Figure 7.1** A possible energy future, targeting reduced reliance on the most costly energy sources.*

If we were to be extremely optimistic, and say solar and wind energy were both to increase their current production levels by factors of 10 in the next 20 years, they would be able to displace approximately 20 percent of today's coal, or 50 percent of global firewood use. However, over those 20 years, global energy demand would also increase. Therefore, even an order of magnitude increase in solar and wind power production would fall short of the modest goal of displacing 50 percent displacement of one, often ignored energy source. In order for solar and wind together to displace half of firewood and half of coal would require each energy system to increase by more than a factor of 50, again, at current levels.

Consider as well that in 2010 the combined solar and wind power production amounted to about 3 Quads out of the 524 Quads of total global energy. If each of them increases by a factor of 10, they would be producing a combined total of 30 Quads, but by 2035, the global demand for energy is expected to increase by more than 250 Quads (EIA 2013). Global demand will increase by five times as much as the growth of the most popular alternatives. The traditional energy sources will need to increase their production until at least the middle of the current century.

Even more aggressive growth in the popular renewables could only reduce a fraction of one of two of the most problematic resources, while not touching crude oil or nuclear, let alone natural gas. The net value of our collective energy systems can increase more rapidly if we make our selections to optimize our use of energy, including shifts to more advantageous fossil fuels.

Additionally, the transition to increased energy value is not restricted to the selection of sources, but should include our consumption patterns as well. Conservation and efficient use are important ways to enhance overall energy system value. Indeed the less intensive solar and wind sources virtually require more efficient uses, to make the limited energy flux go further.

Conservation is basically without costs. It is as close to a free ride as you can get. It may involve some sacrifices, in choosing to forego some activities. It will certainly require some changes in our lifestyles and choices. Many of the more deliberate choices may actually produce benefits of their own, such as more exercise and money savings. Replacing waste with conservation should clearly be one of our top priorities.

The transition to conservation and greater efficiency will have some painful moments, but will ultimately give way to increased value. For instance, more efficient lifestyles, especially in the United States, will entail the demise of exurban sprawl. There will be a great deal of idled housing stock and some economic value will be lost in the process. In the long run, though, it holds the promise of re-building community, in which people have more time at home with family and less "quality" time isolated in automobiles. People may also find that walkable communities add value to their lives. We urge reading one of James Howard Kunstler's books, such as *Home from Nowhere*, for a colorful view of this transition.

In the near term, some of the fossil fuels, most notably natural gas, may add net value to our global energy systems if we increase their use. If that additional use displaces firewood demand or perhaps, coal demand, it will almost certainly mitigate climate change, reduce particulate air pollution, preserve ecosystems, and reduce fatalities. Hard to beat.

Furthermore, although engineered biomass is likely to burn more cleanly than raw biomass, significant safety and health hazards exist. Indeed, the fatalities in acquisition for biomass crops will probably be as severe as for fossil fuels (Sumner and Layde 2009).

Addressing Needs in Poor Regions

Consider, also, that if the world decides to control energy consumption and only used a modest 1 percent per year global growth rate until 2050, it would be minimal to achieve any reasonable access to energy for all human beings. Under those conditions, there will still be some deficit, shown as demand deficit at the top of Figure 7.1. That deficit may be made up by increased natural gas production or by the use of nuclear electric power. This should help clarify that, while we may be able to eliminate the worst energy source and reduce the worst fossil fuel, we cannot expect to eliminate all fossil fuels under any conditions that respect the development needs of the less affluent half of humanity.

Even for those that believe that the overall primary energy demand will decrease due to improved efficiency and conservation, we believe that such assumptions are implausibly optimistic. Half of humanity lacks access to basic modern energy services and the global population, especially in poor regions, is growing. Efficiency and conservation can only partially offset the need for growth.

For the foreseeable future, the values that lifted fossil fuels into market dominance will keep them there. The sheer magnitude of their production represents a huge mountain for the alternatives to climb. Additionally, the extraordinary energy densities and dispatchability offer critical values. The alternatives are far from being able to produce at comparable scales.

Even with continued exponential growth in solar and wind, coupled with renewed development of geothermal power, and a combined oil and gas production peak forestalled until the latter half of the century, a gap is likely to develop between global energy demand and supply.

Shifting Scales

Scales may be changing though. Many energy experts are viewing distributed energy systems as an essential part of the transition to more sustainable systems. Centralized energy systems prevailed in much of 20th-century energy development. They provided economies of scale, which fit with the enormous appetites for energy demonstrated by affluent nations. Rockefeller famously sought to bring order to oil markets, building large refineries, first in Cleveland and pulling a great many strings to bring as much oil as possible through his well-organized system. The large-scale refineries offered the range of products people increasingly desired and tended to concentrate the operations.

Electric power generation is more central to the debate about distributed or centralized energy. The large-scale, centralized systems were able to generate not only vast quantities of power, but the national grids as well, which can direct power when and where it is needed. Or it can trigger widespread disruptions if a component of the interconnected national or regional grid fails.

Today, the World Alliance for Decentralized Energy (WADE 2015) bears the standard for a transition to distributed energy systems. The transition will probably support the rise of the alternatives because decentralization will help to compensate for the infrastructural deficit faced by the alternative systems, compared with the established fossil fuel and nuclear-based energy sources. Indeed, it will be new, localized infrastructural systems, which may help to create a level playing field. However, fossil fuels are highly scalable and can operate in small-scale distributed energy systems too. That's how it all started after all. The Drake Well in Pennsylvania started the first oil boom with a production rate of 10 barrels per day. Distributed, localized energy production will probably be part of a more sustainable energy future, but it doesn't preclude the fossil fuels.

Building a Utilitarian Assessment

Many controversies rage about energy systems and their impacts. Most of them speak to some uncertainties or to some core values that may legitimately lead to different outcomes. Part of the challenge in building a consensus around energy choices during times of transition can be precisely how we value externalities and what values we seek from energy systems. The need to build some consensus leads us to focus on a utilitarian argument.

The purely ethical issues are important, but are difficult to universalize. We all have different values and different belief systems that inform our ethical stances. Many people view the Biblical statement that "Man shall have dominion over the creatures of the earth" to mean that all of the creatures of Earth are only here to serve us and, thus, deserve no more consideration than what they do for us. Conversely, some people hold strong belief in rights of all creatures, a concept advanced as "reverence for life" by Albert Schwietzer. These and many other conflicting belief systems and ethical stances are essentially impossible to reconcile. Therefore, we elect to pursue utilitarian foundations for the analytical paradigms, in order to seek common ground.

There are a few points that we think are universal. First, the true net energy yield is critical to an energy source. If you take a job, the true net income really matters. If you have more expenses than revenues, you cannot make a living with that job. It really does not matter what the expenses are: purchasing supplies, fees, dues, royalties. In terms of whether or not you are or can be making a decent living, is a matter of what the income is, after *all* expenses. The same is true for energy sources. It really matters how much net energy they yield. It is a meaningless accounting exercise to count only the fossil energy inputs. If the total energy value is near zero, then it is a highly dubious energy source, just as a net income near zero is a dubious livelihood.

The same is not true for an energy carrier. A carrier will have a net negative energy balance. The point of using an energy carrier is to enhance the value of the energy consumed: by making it portable, or more versatile, or more powerful. Electricity is extraordinarily versatile. It can support any kind of task and is clearly the only choice for many tasks in the modern world. It can be transported and delivered modest distances with both efficiency and precision, with many devices drawing very tiny currents. This delicate control enables all of modern computing, telecommunications, and electronic entertainment. Lights can be turned on at the flick of a switch—and dimmed. Electricity is also of high value because it can be instantly available at any outlet. It is extraordinarily convenient and effective for all those who have access to it. The value of energy carriers is based on the value enhancement from the original source.

While the marketplace shows some signs of shifting toward the flux-limited, renewable resources, it is, in practice, a minor shift. Take solar PV cells for example. The prices on the manufacture of solar cells have indeed dropped, which have been a boon to the solar installer industry. However, at present, the floor seems to have been reached, and we seriously doubt that installation prices, which account for about half of the total system investment, are likely to drop. More importantly, the sheer magnitude of manufacturing will limit the rate of growth in solar PV production, which should not be surprising (indeed, recent material shortages caused temporary spikes in prices, even at the very limited levels of current global production). Hence, while solar power will most probably become a major contributor to the global energy mix at some point, that point remains decades away. We would suggest the latter half of the 21st century as a reasonable marker.

Wind enjoys a much larger, although still minuscule, global energy market share as of 2015. In certain markets, such as in Germany, we will see wind power representing a substantial percentage of generation at certain times of the year, but we must be mindful of pointing at special cases and extrapolating to larger regions and markets. Wind can grow at a larger absolute rate from its larger base, but evidence suggests that the percentage growth is falling behind that of solar power.

We must also be mindful of context in evaluating the value of energy sources cannot be overlooked. Energy systems must be properly matched with needs. Fundamentally, energy only offers value to the extent that it serves people's needs. To do so, it must work. Locally produced resources will be particularly valuable, especially in Developing Countries. As we enter a time of global energy shortage, the value of energy for its own sake is likely to eclipse the value of cash earned from exports.

Progress toward Maximizing Value

We cannot do without the value provided by energy. Knowing that there is no free ride, that there are costs as well as benefits to every system, we can only plan how to design energy systems that provide the greatest value, with the least costs.

This century will be one of energy transitions, and we can only seek to guide those transitions toward increasingly effective, increasingly clean, and safe energy systems. It will take time and value potentials, for different sources will rise or fall with passing time.

There will be no easy transition. We can expect prices to rise. When crude oil production reaches or even approaches a global peak, prices will rise. People will be distressed. We want our energy cheap and abundant. No matter how much we would like to have all the energy we want cheaply, we should acknowledge both the inevitability and, even the benefits of higher prices. Higher energy prices will:

- Support conservation
- Support the economic competitiveness of alternatives
- Increase our awareness of the values—and costs—of our energy systems.

The transitions must also provide a path away from raw biomass for half of the world that still depends on it. To that end, fossil fuels are alternative energy sources. They will provide greater value than firewood in the Developing World, much as they did for the Industrialized World. In the nearest term, improved cook stoves can reduce the demand for firewood, but the transition cannot stop there. Modern fuels need to be introduced. Over sufficient time, electricity from noncombustion sources may supplant combustion fuels for cooking, but that transition cannot be expected to happen easily. Locally produced natural gas has potential to add tremendous value in this context.

Even coal may offer greater value than firewood, as it did at the dawn of the Industrial Age in England and Western Europe. At the very least, it can save forests from the firewood seller's axe, as it did then and there. Nevertheless, the value offered by coal is severely impaired by the environmental costs in surface mining, the human health costs in underground mining, and the massive emissions from burning it. Globally, we should seek to limit the use of coal. In the short term, this can best be achieved by some combination of natural gas and nuclear power.

Fears of nuclear meltdowns, weapons manufacture, and of radiation in general will limit the growth of nuclear power. The fear that processing facilities for nuclear fuel could also produce weapons grade materials will almost certainly prevent broad dissemination of nuclear technology among the many nations of the world that don't already have a nuclear program. We believe that the nations that already have nuclear technology and no large domestic oil or gas production will continue to develop their nuclear power, affording modest growth in the value added by nuclear power for the next few decades. Once the world is in a global energy shortage, the fears may be swept aside by the urgency of need, but then it will still take a few decades for nuclear power plants to be approved, built, and put on line in vast numbers. It should also be noted that the urgency of need in Developing Countries will still not allay the anxieties in the affluent world about sharing nuclear technology. Its value can be greatly enhanced by making clear, effective decisions about how to handle both high-level radioactive waste and the massive quantities of low-level radiation in decommissioned nuclear power plants.

Oil and gas will continue to grow and to add more energy use value as long as their finite stocks will permit. Gas in particular, offers the most energy value for the lowest negative externalities of any of the combustion fuels. This includes carbon emissions.

The oil peak will most probably occur around the middle of the century. However, the shortage will likely precede that by nearly 20 years, as the rate of growth in petroleum production slows, but demand doesn't. The fact that the gas peak lags behind the oil peak will help to mitigate the shortage. The rising prices for fuel will spur greater exploration for oil and gas and more enhanced recovery technologies for oil. The rising prices, though, will also open the door for the alternatives to compete much more effectively.

Until the global peak in petroleum production, the noncombustion alternatives will struggle to compete with the convenience of highly dispatchable energy sources, with high energy density and well-established infrastructures for delivery and use. At that time, the alternatives with the greatest apparent potential will draw considerable corporate investment.

Before the production of solar or wind power can scale up immensely, their manufacturing will have to scale up just as dramatically. This vast manufacturing will also consume a great deal of energy, which will take years to repay through the energy produced by these new systems. It will probably be mid-century before wind and solar can combine for more than 10% of global energy: longer if oil and gas are able to produce in excess. (Remember that global energy is much greater than global electricity or "power" production, on which wind and solar percentages are often reported.)

Both solar and wind will benefit from being paired with more dispatchable energy sources. With the advent of modern gas turbine technology that can follow load changes, natural gas has the potential to add tremendous value to solar or wind systems. Hydropower and geothermal also have similar potential.

High energy prices will also motivate expansion of other energy sources. Geothermal may be one of the most interesting, as it has a vast, ubiquitous resource base. It can add a great deal of value, as it releases no combustion products and is capable of producing energy at a large scale from individual wells.

Timelines for 21st-Century Energy

Consider Figure 7.2. It represents a simple view of a sunset/sunrise timeline for energy systems. Solid lines represent major production for each resource. Dashed lines represent limited production. Dots represent production rates that are negligible relative to global demand and blank space represents no production. An "X" is placed on the timeline for the most problematic resources (i.e., firewood and coal) to represent plausible times to eliminate or greatly reduce (sunset) use of the resource. Because controlled fusion still requires considerable research before it can even begin to scale up as a commercial producer, it shows no production for the next few decades, followed by a question mark, indicating that we can't know when or even if fusion will rise and become a commercial energy producer.

Whereas this figure only represents a possible set of conditions, it does serve to illustrate a part of the challenge that some resources will diminish soon. Some of them need to be reduced. Nothing is likely to gain large market shares in the near term. The total energy value of the overall energy system must increase, but the time function is one of the important measures that must be included in a viable metric.

Energy decision making will need to be informed by metrics that account for all of the factors that affect the value added by energy systems, including the costs and benefits of energy conversions and end use must be included in the overall metric.

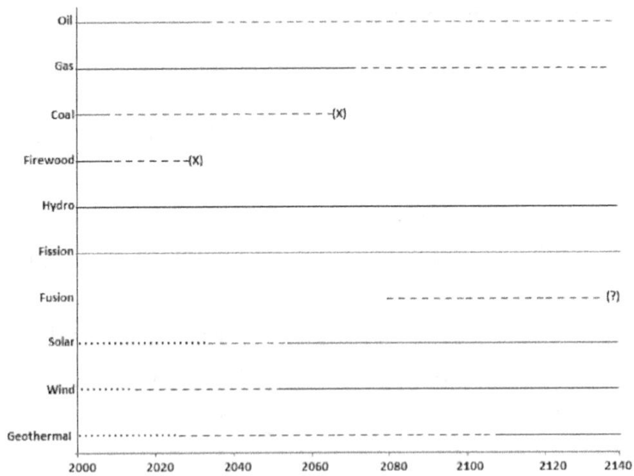

Figure 7.2 A sunset/sunrise timeline of energy resources.

Measuring/Quantifying Energy's Value

We've laid out many of the factors that add to or detract from the value provided by different energy systems. The challenge is to formulate metrics that take these factors into account in some reasonable manner. No metric for energy value can be perfect, in part, because energy systems involve so many varying unknowns. Even more fundamentally, though, a metric seeks to quantify issues, many of which are innately qualitative.

In order to model energy values, we need to be able to assess the truly quantitative factors, such as the quantity of current production of a resource, or the total available resource. Those factors must be combined with some that are ideally quantitative, but whose values are extremely difficult to measure: The health impacts of radiation exposures would be in that category. Estimates must be made. Finally, all of the quantitative factors must be combined with those that are truly qualitative, but for which quantities have been assigned. This ultimate combination provides the overall value estimate.

Several models have been proposed for evaluating energy values (see Inman 2013; Sandia 2009; UEM 2008; Wang 2003). They have a number of strengths on which to build, but we suggest that by their very nature, none of them adequately capture the totality of value costs and benefits that are important in evaluating the contributions that various energy systems can make. Particularly, we see several important points that should be addressed in energy value models:

- They should account for values of the needs met (survival, development, fulfillment, luxury).
- They should account for the value added by the energy in terms of versatility, energy and power density, or dispatchability.
- They should address social costs and benefits.
- They should be based on contextual functions.
- They should include a time function to account for the declining future value of depleting resources and the lag time for new energy systems.

A comprehensive energy value model is quite complex. Some of the current modeling approaches seem to add complexity in their effort to define new terms. We suggest that, whereas the complexity cannot be avoided, the most productive efforts will be those that seek to add clarity. A good energy value model will inevitably be subject to argument, but should be accessible to readers.

Sustainability

One of the key values to be sought in energy systems is increased sustainability. Of course, in order to measure it, we must understand it. Different people use the term sustainability differently in different contexts. Both the profundity and the difficulty of the concept are in its breadth. Some people prefer the term "environmental sustainability," which certainly focuses on one of the important aspects, but does not encompass the totality. Others use it simply to mean that a program or company has plans for their own continuance in the future. That narrow usage lacks all of the robustness of the social, environmental, and resource meanings of sustainability.

The term itself comes from the concept of sustainable development, which is based on meeting the needs of people: currently and in the future. Sustainability depends on sufficient resources to meet the most important needs, through time. Gross inequity, with half of the world's population living in abject poverty is clearly in contrast with this concept. Therefore, true sustainability calls for understanding and addressing needs.

It seems reasonable, then, to suggest that the goal of sustainability is not a singular outcome, but rather a process of moving toward enhanced sustainability. Indeed, it is a progression away from the most unsustainable activities and resources. Each step in the right direction is a part of the transition. There really is no end point. The world itself is not indefinitely sustainable. Limits are likely to be encountered for every energy source and system possible. It may be some unforeseen environmental impact, or some limiting resource to manufacture the means of an otherwise sustainable energy system. There will be costs with everything we do, but we can, and must, take each feasible step to reduce negative impacts and increase net value. Of course, it is more difficult to measure a process than a singular goal.

We suggest that the first essential piece to making effective transitions toward sustainability is to have a good working definition of sustainability. We cannot measure our progress toward an ill-defined goal. The definition must be one that acknowledges both sides of the equation defined by the Brundtland Commission 30 years ago: meeting the needs of people today, while preserving opportunities for the future (WCEE 1987). It is not simply about the environment. It is not simply about protesting anything we don't like. It is certainly not simply about planning for profitability several years down the road. It is certainly not about defining enemies: Big Oil; Big Coal; Nuclear Power; and the 1 percent. The first three popular enemies supply 90 percent of the energy that meets the needs of people today.

We must move beyond the mindset of blaming "others" in order to move toward more sustainable systems. We can and we should address disagreement, but enmity does not build a strong foundation for the goals of sustainability. Indeed, as we look for flaws to challenge, we must look inward. What can we each do to change to live more sustainably? Then what can we urge others to do?

Part of the problem of "other" is also seen in the development framework. Should others, in the Developing World, be allowed to build massive coal fired power plants? That is not the question. The question is what set of energy systems can deliver the greatest net value. Certainly, excessive use of coal, combined with a vast surge in vehicular use in cities like Beijing, is causing egregious pollution. However, the mistake is not in using coal, gasoline, and diesel, but in overusing them. Waiting for enough solar and wind power at the scales needed would have delayed the much needed economic development for more than a billion people by decades. And, of course, the flow-limited resources would not provide the required level of energy production on demand.

Maximizing value is at the heart of sustainability. Unfortunately, people do not always recognize what is in their best interests. If given the option, we will overconsume. How else can we explain the obesity epidemic? We must all adopt new behaviors that acknowledge that there are limits to everything, including the benefits to be derived from more consumption.

Time

Since sustainability is a process, it leads us to discuss time: the dimension that we move through in a set direction at a set pace. Time is always measured. It controls us in many ways, especially if we do not plan for it appropriately. Resources have similarities to the cycle of life, in which people are born, grow to maturity, live productively for some time, and then face inevitable decline. It is so with resources as well.

The specific factors we must consider related to time are:

- Finite, stock-limited resources provide value within a finite time frame.
- Flow-limited sources provide value only during the times when their flow can be substantially tapped.
- Some resources that have some combination of stock and flow characteristics may be used more sustainably, drawing on stocks to meet demand peaks, but allowing replenishment over a reasonable time.
- Time is required for new technologies to be developed, tested, and refined for the applications required.
- Time is required for manufacturing and deployment of new systems.
- Time is required to develop trained and experienced engineers and operators for new systems.
- Time for the transition is limited by peak production constraints and by potentially catastrophic resource impacts—most notably climate change.

Time Limited Values of Finite Resources

In a sense, time is a part of our environment, but one that impacts us, while we have no impact on it. We have already discussed peak oil and peak gas production, which currently provide approximately two-thirds of the energy value in the world today, but whose values are limited in time. The same is true for coal and, ultimately, for nuclear fission materials. Although we cannot know the exact timing (just as we cannot know the span or our lives), we can and do know that continued use depletes the stocks from which they are each drawn.

The production rates for each can continue to rise until the remaining stocks will no longer support growth in production. At some point, though, when the global production rate is very high, the peak will be reached. Beyond that, production may be able to level off and hold reasonably steady for some years, but a decline is inevitable. Of course, all of the fossil fuels share the characteristic of stock-limited resources, as does nuclear fission. Depletion over time is what most clearly limits the values of these resources.

Biomass is like fossil fuels in being stock limited and potentially depleted. The difference lies in the fact that the growth, or renewal rate, for current biota has the potential to replenish even depleted stocks in an observable time frame. That is, only if the consumption demand is removed or reduced below the growth rate of the biota. Thus, it has some characteristics of a flow-limited, renewable resource, as well as those of a stock-limited, nonrenewable one. Nevertheless, when its use exceeds its regrowth rate, the depletion harms ecosystems and the future growth of the resource. Primary use of biomass as an energy modern source (other than waste streams) is probably unsustainable.

Time Limited Value of Flow-Based Resources

Solar and wind are the two resource systems that truly rely exclusively on tapping the natural flow of energy. Their use is limited by the variable flux of the energy stream. For solar, there is a predictable variability, based on day and night. For both wind and solar, there is a less predictable variation based on weather patterns. Their availability is generally accounted for by the term "load factor."

Another important factor is the time that will be required for some of the energy systems to grow from extremely limited starting points. These are not like wells, which can produce vast quantities of oil or gas, simply flowing from the wellbore. Solar and wind energy conversion devices must be manufactured. Particularly in the case of solar PV cells, each manufactured device produces very little energy.

Consider the notion that solar PV cells could provide 20 percent of the world's energy within the next 20 years. If annual world energy consumption persists at 15 TW (15×10^{12} W) and we assume that an individual panel is rated at 200 W and operates at a load factor of 0.2 (i.e., 0.2 W are produced per watt installed due to weather, climate, and geographic factors), then the number of panels needed to be manufactured per year to meet this goal is calculated as:

$$\frac{15\times10^{12}\ \text{W global power}\times\ 20\%\ \text{of global power}}{200\ \text{W/panel}\times 20\,\text{years}\times 0.2\ \text{W produced/W installed}} = 3.75\,\text{billion panels / year}$$

3.75 billion panels per year, for 20 years. That equates to over 10 million panels per day or almost 120 panels per second at continuous production. Remember that this level of production cannot even start until the factories are built… and the mines… and the infrastructure for delivering raw materials to the factories.

Very recently, some large manufacturing has gotten underway. What appears to be the largest PV manufacturing entity in the world reached a production of almost 2.5 gigawatts (GW) of capacity in 2014 (Yingli 2014). That equates to one large supplier achieving well under one-one-hundredth of a percent of the required manufacturing. So, while the goal may be possible, it is optimistic and very far from where we are right now. Remember, as well, that the 20-year clock doesn't really start until sufficient global manufacturing is in place.

How much manufacturing capacity must be built to accommodate this? How much impact will there be in mining all of the materials for this and in all of these massive manufacturing activities? Will there be accidents installing 10 million solar panels every day?

Then, the activity must continue at this extreme pace if solar PVs are to maintain and, ideally, increase their market share. Today's PV manufacturers are suggesting average lifespans of 30–40 years. Therefore, a great deal of the manufacturing activity will need to continue just to replace systems at the ends of their lives.

However, global demand will probably grow until most people have access to adequate modern energy and until populations stabilize. Until then, PV manufacturing will have to keep up a frantic pace, just to secure a modest market share: one that could displace firewood demand and about half of coal's market share. Remember that this does not truly account for growth in demand over the next few decades.

If wind also continues dramatic growth, similar manufacturing will be required. Its growth will be further limited by the availability of prime wind sites reasonably near population centers. So far, it has grown rapidly, partly because many ideal sites were available, but now they are largely taken.

Together, accounting for growth in demand, it seems quite optimistic to expect wind and solar to secure 20 percent of the global energy market. This market share will be significant and can achieve some added value. If it targets replacing the worst energy sources, though, it will not have any significant impact on the less polluting oil and gas production, or nuclear power.

Innovation Time

In some ways, time will almost certainly lead to innovations that solve most of the problems we are dealing with currently. Human creativity is, for practical purposes, an unlimited resource, but it takes time. Creative thinking takes time for research, reflection, and testing. We discussed in the first chapter the notion that one of the great advantages that fire offered early humans was the time to create: expanding the working day after dark.

Some people would raise the point that the time for creative development is a function of researcher days, not merely days. There is truth in that. The more scientists and engineers working on problems, the more rapidly innovations will arise. The caveat to that point, though, is that research is an exploratory process, not a simple journey to a designated end point. (If we knew precisely where we were headed, research would not be necessary.) Thus, the many innovations developed by many researchers cannot be counted on to meet specific energy needs.

Innovation through Scale-up to the Marketplace

Scale-up, taking technologies from the lab, to a pilot project, and on to full, commercial scale implementation is an involved process, in which many ideas fall by the wayside. Failure to reach commercial scale is a part of the (almost) natural selection process. It weeds out many flawed concepts, but it can strike down potentially valuable ideas too. Recall that the marketplace is not an omniscient divinity. It is a set of forces that tend, in general, to drive innovation and improvements. Unfortunately, its actions are based on the perceptions of consumers and investors: perceptions that can be skewed by misinformation, prejudices, or lack of understanding and imagination.

That is one of the purposes of writing this book, to try to give readers a few additional tools to improve their perceptions when viewing energy-related technologies. If a resource or technology is perceived as good and promising, it will continue to draw investment, even if it is fatally flawed. Our modern society, with a range of competing special interests, lobby groups, lobbyists, and advertisers, can provide a wide range of misinformation.

However, the competing interests will balance each other out, yielding good information, right? Generally not. Most of us tend to listen to or read those sources that support our preconceived notions, our political, leanings, or our own special interests. Even we are not immune to biases. Fortunately, as two very different people working together on the book we are forced to pay attention to each other's perspectives and it does help to create some balance. That cannot eliminate all the biases that can mislead, so we also try to make our analyses, arguments, and assumptions as transparent as possible, so that when you disagree with a conclusion, you have a decent basis for your disagreement.

Most new ideas have to fight their way through biases to achieve attention, and any opportunity to move toward scaling up production. Even before trying to convince others of the importance of a discovery, a serious scientist or researcher approaches his or her own work with some skepticism.

Final Thoughts

We have attempted to develop a case for the creation of a new metric, which accounts for the value added by energy as well as its costs. We suggest that such a metric must be dynamic across time. Some resources will deplete, whereas other energy systems require time to scale up, and yet others require research and development to reach a viable technological maturity. Such an encompassing metric will be replete with assumptions and approximations, which will need continual review and revision.

There is a great deal we cannot know. How much oil does the earth really contain? How long will it be before nuclear fusion is technically and commercially successful? Indeed, will it ever be? These and many other unknowns will vex energy systems analysts and the people who ultimately need to be informed by them.

Most serious decision-making operates within uncertainty. If the performance of a given stock, or the stock market in general, were knowable, then there would be no financial collapses. Individuals would need only to find whose advice they should follow in order to become wealthy. Of course, if such knowledge were there for the taking, no one would invest in the wrong companies or technologies and competition would rapidly dry up.

But there are so many decisions that we make in uncertainty. Where should I go to college and what should I study? Should I marry this person? Should I start writing another book? We cannot know the answers to any of these questions. Yet we must make decisions. Even though we cannot know all the answers precisely, we can limit the range of answers by what we do know.

Consider the great, typically unspoken question: how long will I live? Barring a fatal disease with a very short prognosis, none of us can know the answer until the moment is at hand. Then it is generally too late to make any plans. Nevertheless there is something we can know about the question. We're going to die. No exceptions; everybody does. We can even know some reasonable ranges.

Very few of us will live more than a century. True, there are those claiming that medical advances that will greatly extend the upper limits of life are just around the corner. For any of you who actually believe you will live to be 200, the only thing stopping us from offering you a wager is the likelihood that neither of us will be around to collect. We can know with reasonable certainty that none of us will live much past 100 years, and most of us will die well short of that mark. We can't know precisely, but there are limits, or bounds, of reasonability that we can assign to the question of our longevities.

This is how we propose to construct a viable metric. Begin by defining the limits within which system values must be constrained. Finite resources cannot last indefinitely. New technologies cannot be broadly disseminated overnight.

Realistically, we would suggest that the fossil fuels will be peaking and beginning a decline by the latter half of the current century. Their finite occurrence will most likely be their limiting factor. Until those limits are felt, the versatility, dispatchability, and low economic cost will make the fossil fuels difficult to displace.

Nuclear fusion will not be a viable energy source until at least the time when crude oil production peaks. Fusion will not add any value before that time. In the metaphoric tree of energy sources, it is merely a seed. It may grow to be a tree of its own, but for now it is merely a dormant possibility for the relatively distant future. Until then, fusion research consumes energy, without producing any net useful energy.

Nuclear fission, on the other hand, is a proven technology that utilizes a very large resource base. It can grow. It requires massive capital construction, which takes years for each new power plant. But, it won't take off in many countries as long as fears persist. Its near term future is a question. In the longer term, it is likely that energy shortages will help to overcome fears and it will grow.

The popular noncombustion alternatives will also grow. Their growth is limited by the small base from with they are launching. It will take considerable time to reach a large enough production base for doubling to mean much in terms of absolute energy production.

Time is one resource in short supply for transitions. Oil gas and coal are finite. Crude oil will reach a peak, likely in the next few decades. Natural gas will lag a few decades behind, but still reach a global peak in production, probably within the current century.

At the same time, some energy sources should be eliminated, or at least reduced, as soon as possible. Raw biomass dependence should be at the top of that list. Besides the enormous cost of human lives, it entails serious environmental impact, and offers very little value. Coal is clearly the most socially and environmentally costly of the fossil fuels. In spite of the vast resource base, coal should be curtailed as quickly as possible.

Scoring Resources at a Rudimentary Level

Here are some of the most important points we'd like to leave you with about energy values:

- Energy is essential—we can talk about using it more efficiently, but cannot do without it.
- Energy systems have generally developed because of the value they add.
- The fossil fuels and nuclear power produce too much energy to be replaced easily.
- Raw biomass dependence is the most important energy source from which to transition: It incurs by far the highest cost to human lives and safety, while also destroying forests and not supporting development.
- The next largest cost to human safety comes from individual transportation. For the sake of safety, of efficiency, of animal lives, and of reducing emissions, we must transition to more modern public transportation and to living closer to our work and activities. We cannot expect to eliminate private transportation, but it should be reduced as much as is practical.
- No transition will occur quickly or easily.
- Climate change is an important challenge for humanity, but information about it does not seem to be fundamentally altering human behavior.
- Price is the most likely driver of fundamental change.
- Every human activity has environmental and social impacts; no energy system will be perfect.

Hence, every energy system has costs and benefits and limiting factors. The decisions we make for a future transition to cleaner, more equitable energy choices must be guided by at least some sort of contextual decision matrix. How do we even begin to make these types of decisions in a fashion that goes beyond qualitative observations and opinions? Metrics surely help pave the way for rigorous approaches to answering energy transitions questions, but ultimately, decisions must also include an informed balancing of costs and benefits that the citizenry will accept.

Below is Table 7.1 that we think helps to summarize what can be said today about overall energy values of various systems. We opted for a very simplified scoring system, so that the reader can see a numerical expression of what we offered throughout the book, albeit extremely unsophisticated. The simple quantification process that we used was to apply the following number scores to these qualitative assessments: –3 for extremely negative (assigned to the health impacts of raw biomass, because the death toll is monstrously large); –2 for very negative; –1 for negative; 0 for neutral; +1 for positive; and +2 for very positive.

A great deal of work has yet to be done, but, even so, we suggest that this simplistic approach highlights the resources that need to be curtailed as quickly as possible: fuelwood dependence and then, extensive coal use. The world faces an enormous challenge to increase the true, net value we receive globally from our energy systems. There can be no doubt that energy access must increase to serve the needs of more than 3 billion people who lack adequate modern energy. The costs to people's lives and to the environment must be reduced at the same time. This will require a rapid transition away from unsustainable fuelwood dependence. The challenge will be immense. We can only hope to address it effectively by understanding both the values and the costs from each energy system. Best of luck to all of us.

Table 7.1 Energy system scoring scheme to assess overall values to resources.

Resource	Abundance	Safety costs	Environmental costs	Economic value	Social value	Overall value
Raw biomass	Score: –1 Both stock and flow limited. Fuelwood dependence is exceeding growth rates.	Score: –3 Very large—by far the most fatalities of any system.	Score: –2 Smoke, loss of habitat, loss of carbon sinks	Score: 0 Small—some firewood and charcoal businesses, but costs to consumers are high	Score: 0 Minimal.	Total Score: –6 Very small—it serves essential needs, but at huge costs. Should be targeted for elimination
Coal	Score: +1 Finite, but very large stock	Score: –2 By far the most dangerous of current modern sources	Score: –2 Disruption of surface; contamination of water; smoke; largest climate change contributor.	Score: +1 Moderately large. It provides the largest share of electric power generation.	Score: 0 Mixed. Electricity is important to quality of modern life, but the jobs have been hazardous	Score: –2 Small. It serves an important need, but at very large environmental and health costs. Should probably be phased out, but cannot be replaced completely anytime soon.
Oil	Score: 0 Finite resource, but larger than most have thought. Largest energy producer of modern times.	Score: 0 Hazardous work but involves few workers relative to energy produced.	Score: –1 Large spills can have disastrous local environmental impact. Major contributor to climate change.	Score: +2 The largest provider of energy. Highly dispatchable.	Score: 0 Suffers from popular perception problems, yet people continue to consume it heavily.	Score: +1 There are clear costs to excessive oil use, but it provides a great deal of value. It will deplete and should be conserved.

Resource	Abundance	Safety costs	Environmental costs	Economic value	Social value	Overall value
Natural gas	Score: +1 Finite resource, but larger than most have thought. Its use lags behind oil, so should not be constrained until after mid-century.	Score: +1 Hazardous work, but very few workers relative to energy produced.	Score: 0 Contributes to Climate Change, but less so than other combustion fuels.	Score: +2 Adds a great deal of value to the economy, where available.	Score: +1 Suffers from some popular perception problems. Otherwise, its versatility and clean burn are valuable.	Score: +5 Gas can and will offer considerable value. It is a finite resource that should be conserved—flaring should be targeted for prompt elimination.
Nuclear fission	Score: +2 Finite, but enormous resource.	Score: +1 Hazards are greatly exaggerated, but real.	Score: 0 No combustion products. Disposal of radioactive waste.	Score: 0 Requires large capital investment, but can provide significant energy at low cost.	Score: –1 Suffers from public fears. Probably limited deployment in Developing World.	Score: +2 Large potential and will probably need to increase in its use to achieve goals of reducing the worst resources and offsetting oil and gas depletion.
Nuclear fusion	Score: 0 Huge resource base, but no production established.	Score: +1 Probably low.	Score: + 1 Probably low.	Score: –1 None thus far—only costs related to research.	Score: –1 None so far.	Score: 0 It has great potential, if it works. Deserves research, but cannot be counted on.
Hydropower	Score: +2 Large resource base, largest current “renewable” source.	Score: +2 Minimal hazards.	Score: –1 Disruption of ecosystems.	Score +1 Large capital investment, but cost-effective over time.	Score: –1 Mixed. The reservoir may add recreational value, but new dams can displace many.	Score: +3 Large potential, but many of best sites are taken and large dams have impacts.

Resource	Abundance	Safety costs	Environmental costs	Economic value	Social value	Overall value
Solar	Score: +1 Vast resource base, but very little current production.	Score: +1 Few known hazards, but large scale mining and manufacture will have impacts.	Score: +1 Probably small, but large scale may reveal impacts.	Score: –1 Currently becoming competitive, based on direct subsidies.	Score: +1 Good perception, but limited on-demand capacity.	Score: +3 Doubtless an important resource for the future, but very limited value added at present and unknown costs at large scale.
Wind	Score: +1 Very large resource base, but small current production.	Score: 0 Large-scale wind turbines require massive materiel and high tower work.	Score: +1 Concern for bird and bat kills may be exaggerated and noise pollution concerns are extremely site specific.	Score: 0 Large capital costs, but competing currently, based on direct subsidies.	Score: +1 Popularity fading as it moves to larger scale.	Score: +3 Will probably continue to increase the value added, but at a much slower rate.
Geothermal	Score: 0 Very large resource base, but small current production.	Score: +1 Minimal risks, related to drilling and to handling, hot, high-pressure fluids.	Score: +1 Closed loop systems have virtually no emissions, some land use.	Score: 0 Economic in limited high-pressure, high-temperature applications.	Score: +1 Relatively unknown, but provides electricity on demand.	Score: +3 Very clean, but depends on technologic advance to tap lower enthalpy resources or Hot Dry Rock.

References

Energy Information Administration of the US Department of Energy (EIA). 2013. Annual Energy Outlook 2013.

Inman, M. 2013. "How to measure the true cost of fossil fuels." Scientific American, 308(4), 58–61.

Sandia (Sandia National Laboratories). 2009. "Models for evaluation of energy technology and policy options to maximize low carbon source penetration in the United States Energy Supply." Sandia Report SAND2009-8205. Accessed from: http://prod.sandia.gov/techlib/access-control.cgi/2009/098205.pdf

Sumner, S., and Layde, P. 2009. "Expansion of renewable energy industries and implications for occupational health." JAMA, 302, 787–789. Accessed from: www.jama.com.

Urban Environmental Management (UEM). 2008. "Life cycle analysis and assessment." Accessed from: http://www.gdrc.org/uem/lca/life-cycle.html

World Alliance for Decentralized Energy (WADE). 2015. "What is DE?" Accessed from: http://www.localpower.org/deb_what.html

World Commission on Energy and Environment (WCEE). 1987. Our Common Future. New York: Oxford University Press.

Wang, M. 2003. "Well-to-Wheels Energy and Emission Impacts of Vehicle/Fuel Systems Development and Applications of the GREET Model." Center for Transportation Research, Argonne National Laboratories. Accessed from: http://www.transportation.anl.gov/pdfs/TA/273.pdf

Yingli. 2014. "About Yingli." Accessed from: http://www.yinglisolar.com/us/about/

Index

OTHER TITLES IN OUR ENVIRONMENTAL ENGINEERING COLLECTION

Francis Hopcroft, Editor

- *Wastewater Treatment Concepts and Practices* by Francis Hopcroft
- *Small-Scale Wind Power: Design, Analysis, and Environmental Impacts* by John P. Abraham and Brian Plourde

FORTHCOMING IN THIS COLLECTION

- *Hydraulic Engineering: Fundamental Concepts* by Gautham P. Das
- *Chemistry for Environmental Engineering* by Armen S. Casparian
- *Geotechnical Site Characterization* by Anirban De

www.ingramcontent.com/pod-product-compliance
Lightning Source LLC
LaVergne TN
LVHW052316200926
848481LV00002B/578
9781606502631